Applications of Tree-ring Studies

Current Research in Dendrochronology and Related Subjects

edited by

R. G. W. Ward

BAR International Series 333
1987

B.A.R.

5, Centremead, Osney Mead, Oxford OX2 0DQ, England.

GENERAL EDITORS

A.R. Hands, B.Sc., M.A., D.Phil.
D.R. Walker, M.A.

BAR -S333,1987:'Applications of Tree-ring Studies'

© The Individual Authors,1987

The authors' moral rights under the 1988 UK Copyright,
Designs and Patents Act are hereby expressly asserted.

All rights reserved. No part of this work may be copied, reproduced, stored, sold, distributed, scanned, saved in any form of digital format or transmitted in any form digitally, without the written permission of the Publisher.

ISBN 9780860544289 paperback
ISBN 9781407346052 e-book
DOI https://doi.org/10.30861/9780860544289
A catalogue record for this book is available from the British Library
This book is available at www.barpublishing.com

Contents

		Page Nos.
Growth Rings in the Aerial and Root Xylem of some North Temperate Hardwoods	P. Gasson	1-19
The Palaeoclimatic Significance of Growth Rings in Late Jurassic/Early Cretaceous Fossil Wood from Southern England	J.E. Francis	21-36
The Contribution of Growth Ring Studies to the Reconstruction of Past Climates	G.T. Creber & W.G. Chaloner	37-67
Relationships between British Climate and the Radial Growth of Quercus Species - an Empirical Approach to Climate Reconstruction	K.R. Briffa	69-90
Dendroclimatology of Pinus sylvestris L. in the Scottish Highlands	M.K. Hughes	91-106
Dendroglaciological Investigations in Norway	J.L. Innes	107-120
Mediaeval Dendrochronology in Exeter and its Environs	J. Hillam & C.M. Mills	121-125
A 700 Year Dating Chronology for Northern France	J.R. Pilcher	127-139
Problems of Dating and Interpreting Results from Archaeological Timbers	J. Hillam	141-155
The Belfast CROS Program - Some Observations	J.R. Pilcher & M.G.L. Baillie	157-163
Sapwood Estimates and the Dating of Short Ring Sequences	J. Hillam, R.A. Morgan & I. Tyers	165-185
A Review of the Methodology for Calibrating Radiocarbon Dates into Historical Ages	T.C. Aitchison & E.M. Scott	187-201
The Belfast 'Long Chronology' Project	M.G.L. Baillie & J.R. Pilcher	203-214
Dendrochronological Studies of Bog Pine from the Rannoch Moor Area, Western Scotland	R.G.W. Ward, B.A. Haggart & M.C. Bridge	215-225
The Dendrochronological Study of Sub-Fossil Wood in New Zealand	M.C. Bridge	227-232

Preface

In May 1984 a one day symposium on the general subject of tree-rings was held at the City of London Polytechnic. The purpose of the meeting was to bring together people working on some aspect of tree-rings, be they in a Department of Botany, Biology, Forestry, Archaeology, Ecology, Mathematics and so on. In fact, the range of disciplines represented at the meeting was wider than anticipated, a testimony to the great range of applications that have been found for tree-ring studies.

This collection of articles is the outcome of that meeting: some of the papers were presented at the symposium and others were written specifically for this volume. The balance of those included here simply represents those received, and does not reflect the much greater balance of the papers presented at the symposium, at which sessions were divided fairly equally amongst the biologists, the archaeologists, the dendroclimatologists and the statisticians. The absence of articles describing the statistical properties of ring-width series in this volume is particularly regretted. Nevertheless, hopefully the papers presented here will convey something of the intention and scope of the May meeting, and demonstrate once more the potential of tree-ring studies in so many fields.

The job of editing a collection such as this is not exactly preferable to a holiday cruising round the Greek Islands, (or indeed, anywhere else!). However, it has been made much easier by the patient co-operation of all the contributors, the reviewers, the publishers and the cartographic staff of the Environmental Studies and Geography Department of the Roehampton Institute. I extend my thanks to all of them.

R.G.W. Ward
Feb. 1987

Growth Rings in the Aerial and Root Xylem of Some North
Temperate Hardwoods

P. Gasson
Jodrell Laboratory
Royal Botanic Gardens
Kew, Richmond
Surrey

ABSTRACT

The woody parts of a tree can be divided into aerial xylem and root xylem. Trunk and branch xylem in a given species usually have similar cell distribution patterns whereas roots are often anatomically different. In Pedunculate Oak (Quercus robur L.) the aerial xylem is ring porous and the roots are either radial porous or diffuse porous. The roots of all species of oak usually have xylem with indistinct growth ring boundaries. In contrast, in European Beech (Fagus sylvatica L.) the xylem in all parts of the tree is diffuse porous or semi-ring porous and has well defined growth ring boundaries.

The differences between aerial and root xylem in oak and beech are shown with particular reference to how well the growth rings are defined, and two other British tree species with ring porous aerial xylem, European Ash (Fraxinus excelsior L.) and English Elm (Ulmus procera Salisb.) are illustrated for comparison. The poorly defined growth rings in many root samples of these and other species (particularly those with ring porous aerial xylem) will often result in difficulties with their dating and aging.

INTRODUCTION

Dendrochronology relies on the recognition of annual growth rings in woods under investigation. In most north temperate species growth rings are readily identified in aerial (trunk and branch) xylem, but they may be difficult or impossible to distinguish in root xylem of the same species. Most tree ring studies are concerned with aerial xylem, and roots are unlikely to be encountered except in bog deposits of natural origin or perhaps as fragmentary material from archaeological sites. However, a knowledge of the anatomical appearance of the roots of common dendrochronological subjects is clearly of practical value when accurate identification of samples is necessary, even if their dating and aging is not intended or possible. This paper illustrates the anatomical differences between aerial xylem and root xylem with regard to both vessel distribution and the occurrence, appearance and relative definition of growth rings. The species selected are Pedunculate Oak (Quercus robur L.), a frequent subject of dendrochronological investigations, European Ash (Fraxinus excelsior L.) and English Elm (Ulmus procera Salisb.). With the exception of beech, in all of these species the root structure is markedly different in appearance from that of the aerial xylem.

CHARACTERISTICS OF AERIAL AND ROOT XYLEM

1: Pedunculate Oak

The aerial xylem anatomy of this and related oak species is well known, and there are many published illustrations and descriptions of "typical" transverse sections of the wood (e.g. Schweingruber 1978; Baillie 1982). The aerial xylem is ring porous i.e. the earlywood vessels formed in the early part of the growing season are much wider than the latewood vessels in a given growth ring. This applies to all aerial xylem formed from the second year onwards in a stem or branch. However, in the first twenty years or more the mean diameters of the earlywood vessels increase annually in a linear fashion and the size difference between earlywood and latewood vessels is less marked in the young tree (Gasson 1984).

Growth ring boundaries are well defined in the aerial xylem of this species, and the ring porous pattern of vessel distribution is apparent in all but the narrowest rings. This pattern has not developed in some rings of the sample shown in Plate 1:1, in which there has been a period of very slow radial growth lasting for approximately 8 years. This sample with unusually narrow growth rings has been deliberately chosen to show an extreme modification of the ring porous structure of the aerial xylem. Variation in ring width in this species of oak is considerable, ranging from 200-8,000 um (8 mm) in material examined by the author (Gasson 1984), whilst it hardly exceeds 4 mm in the Mediaeval and modern timbers discussed by Fletcher (1974). In the period of very slow radial growth shown in Plate 1:1, the boundaries of the growth rings follow the contours of the earlywood vessels of the previous ring, and there is virtually no latewood, making accurate ring counting difficult. A much less common modification in aerial xylem structure was reported by Fletcher (1975), who found that 0.4% of 24,000 rings had abnormally narrow earlywood vessels although the growth rings of which they were a part were of normal size. These abnormalities were found in wood used for panel boards between 1450-1620 to support portrait paintings in England and Flanders, a Mediaeval cupboard from Oxford, and parts of a Roman quay in London. Fletcher suggested that exceptionally cold winters and springs could have caused this abnormality by interfering with cambial activity and cell enlargement during earlywood formation. No such abnormalities were found in modern material by Fletcher (1974, 1975) or by the present author, even in the years following the extreme winters of 1946-7 and 1962-63.

Root structure differs in a number of respects from that of aerial xylem. Growth ring boundaries in the three examples shown in Plates 1:2-4 are all less well defined, and the xylem is diffuse porous (Plate 1:2) or radial porous (Plates 1:3-4). In most roots the vessels are much wider than those of aerial xylem of the same cambial age. In Plate 1:2 which shows a root from a mature tree, with a cambial age of about 10 years the vessels are of comparable width to the earlywood vessels of most mature aerial xylem samples, (cf. plate 1:1, which shows aerial xylem with a cambial age of about 100 years). The vessels are crowded and the indistinct growth rings are narrow. The xylem can be described as diffuse porous. In Plates 1:3 and 1:4 again showing roots with a cambial age of about 10 years, the vessels are considerably narrower and are arranged in undulating radial lines, hence the term radial porous. A very gradual reduction in vessel diameter is apparent across the growth rings, which are reasonably well defined in the example in Plate 1:3, but in 1:4, a tap root, it is virtually impossible to detect ring boundaries. The vessels in these two samples may be narrower than those in the lateral root shown in Plate 1:2 because they are from immature trees. The aerial xylem in these young trees has corresponding narrower vessels. Many lateral and tap roots have less well defined ring boundaries than those shown in Plates 1:2 and 1:3.

The appearance of growth rings at relatively low magnifications has been discussed above. At higher magnification, individual fibres, axial parenchyma and ray cells are more easily distinguished, and growth ring boundaries in aerial xylem are less apparent. Plate 2:4 shows a ring boundary in aerial xylem, in which a wide earlywood vessel occurs within 3 cells of the boundary. The last-formed latewood cells of the preceding growth ring consist of fibres which have the same tangential diameter as the previous and subsequent cells in the same radial file but have a radially "flattened" appearance. In this example the narrow band is only 2-3 cells wide, and consists of normal fibres, whereas in many aerial xylem ring boundaries gelatinous fibres may also be present or predominant. In roots, radial "flattening" of fibres is less apparent at ring boundaries, and the lack of wide vessels at the beginning of a year's radial increment means that growth rings are less clearly defined. Gelatinous fibres are absent from most roots.

In summary, the growth rings of pedunculate oak are well defined in aerial xylem and less well defined or sometimes impossible to distinguish in roots.

2: European Beech

This species has diffuse porous or semi-ring porous xylem throughout the tree. A typical section of trunk xylem is shown in Plate 2:3 at the same magnification as Plate 1. The vessels are narrower and more crowded than in pedunculate oak and growth rings can be clearly distinguished. Plate 2:1 shows an enlarged part of Plate 2:3 in which the fibres at the end of the ring are mainly gelatinous.

In roots, growth ring boundaries are not quite so distinct but they are generally much better defined than in oak (Plate 2:2). As with oak, gelatinous fibres were not observed in roots, although they are of frequent occurrence in aerial xylem. The vessels tend gradually to decrease in diameter from earlywood to latewood. However, vessel width is not as good an indicator of the position of ring boundaries as it is in the aerial xylem of oak. A better indicator in beech is that rays often bulge tangentially at ring boundaries, i.e. they are noded (Plate 2:2 and 2:3).

In summary, the differences between aerial and root xylem in beech are relatively slight and ring boundaries are always clearly discernible. Aging and dating of beech growth rings is therefore usually possible.

3: European Ash

In this species the aerial xylem is ring porous and the roots are more or less diffuse porous. Plates 3 and 4 show that growth ring boundaries are usually distinct, and as in oak they are readily observed at low magnification. As in oak aerial xylem, narrow growth rings consist mainly of earlywood (Plate 3:2). The ring boundaries in aerial xylem are composed of fibres and sometimes vessels and axial parenchyma cells, so they are more complex than those of oak. Each boundary is up to about 4 cells wide (radially) and as in oak and beech the cells in a given radial file have similar tangential diameters but the boundary latewood cells have smaller radial diameters giving them a "flattened" appearance (Plates 3:1 and 3:2). Occasionally a radial pair of vessels consists of one latewood vessel and one earlywood vessel from the following year. This does not occur in oak where the boundary consists only of fibres and ray cells.

The fibres in roots are usually thinner-walled than those of aerial xylem, and it is unclear in transverse section whether some cells are axial parenchyma or fibres. Ring boundaries can be seen in all the roots examined (Plate 3:3 and 3:4, Plate 4), but at the higher magnification (Plate 3:4 in particular) the cells in the boundaries are not markedly different in shape from those formed previously and subsequently.

The relationship between growth rings and annual radial growth is not clear in this species where several roots (Plate 4) have growth rings which merge or divide. Growth rings throughout the tree are usually accentuated by the presence of earlywood vessels very close to the beginning of a given ring. These vessels are usually larger than latewood vessels in the previous ring, but in some roots all vessels are very narrow and sparsely distributed (Plates 4:1 and 4:3).

In summary, growth rings are clearly defined in both aerial and root xylem, but in roots the cells at growth ring boundaries are not markedly different from those in the radially adjacent latewood and earlywood, and growth rings are often discontinuous.

4: English Elm

The English Elm is one of three British elm species which all have similar xylem anatomy. The aerial xylem is ring porous and the roots are diffuse porous. The appearance of the aerial xylem is very distinctive, with groups of small latewood vessels in discontinuous bands alternating with fibre blocks (Plate 5:1). However, this pattern is less apparent in roots where the growth rings are less obvious and the latewood vessels are wider than those in aerial xylem of comparable cambial age. The growth ring boundaries of roots (Plates 5:2 and 5:3) are often poorly defined making it difficult to distinguish between individual growth rings. Plate 6:1 shows a ring boundary in aerial xylem in which a "flattening" of the fibres at the end of the latewood is not very marked. The boundary is made distinct mainly by the presence of very wide earlywood vessels. The vessels at the beginning of some growth rings in roots are often wider than those in the previous latewood which accentuates the ring boundary, and in Plate 6:2 the boundary is very well defined for a root. A probable ring boundary shown in Plate 6:3 is apparent only because of a change in vessel diameter.

In summary, growth rings are well defined in the aerial xylem and less apparent in roots.

DISCUSSION

The anatomical differences between aerial and root xylem with regard to growth rings and vessel distribution are the two characters most useful to dendrochronologists. These and other characters listed by Fayle (1968) and Cutler (1976) indicate that the anatomical differences exhibited by aerial and root xylem are a reflection of their respective environments and physiological roles. The aerial xylem supports the leaves in the optimum position for intercepting light and supplies these leaves with water and nutrients from the roots. The whole aerial system of the tree is subject to variable weather conditions and fluctuations in environmental stimuli whereas the roots are protected and supported by the soil. The roots act as an anchor and as storage organs and supply water and nutrients from the soil to the aerial parts of the tree. Roots are subject to less extreme variations in environment and tend to be less hardy in cold conditions (Kramer and Kozlowski 1979).

The greater uniformity of the root environment may help to explain the poorly defined growth rings in many roots. Conditions suitable for radial growth may change more gradually than in the aerial environment and as a result the radially "flattened" cells at the ends of growth rings in aerial xylem are less likely to develop in roots. Lebedenko (1962) stated that radial growth in roots often extends well beyond leaf fall by which time radial growth in the stem has ended for the season, and Longman and Coutts (1974) gave the seasonal periods of radial growth in the two British Oaks (Q. robur and Q. petraea) as mid-April to mid-September in the stem and May to October in roots. In addition to this evidence by implication, Lohr (1969) found that low light levels induced oak and ash to produce aerial xylem with poorly defined growth rings.

Several authors have also provided evidence that ring porosity in aerial xylem can be modified by environmental stimuli. Knowlson (1939) examined some sessile oak (Q. petraea) trunks that had been buried for several years, during which period they had formed root-like xylem. Conversely, Lebedenko (1962) found that chestnut (Castanea sativa) roots formed stem-like xylem during a period of exposure to the air.

The modification of vessel distribution in roots is linked with loss of growth ring distinctiveness. All three ring porous species examined (oak, ash and elm) have roots which are not ring porous, probably as a result of the more uniform environment of the soil. This loss of ring porosity corresponds with an increase in vessel diameter with respect to cambial age. Roots are closer to the water source of the tree than the aerial xylem, and there is therefore less risk of air embolisms in root vessels in all but the worst drought conditions.

The reasons for anatomical differences between aerial and root xylem are complex. These differences are most apparent in ring porous species and in contrast the only diffuse porous species studied, beech, displays relatively little anatomical variation throughout the tree, growth ring boundaries being well defined in roots and aerial xylem. This is also true of some other diffuse porous species, but lime (Tilia americana) roots examined by Fayle (1968) were variable in this regard. Even in beech, transverse sections at high magnification (Plate 2:2) show that growth ring boundaries are slightly less well defined in roots.

Differences between aerial and root xylem are not restricted to growth rings, vessel distribution and vessel diameter. The storage function of roots is reflected in a higher parenchyma and correspondingly lower fibre content in most roots, which was observed in all the species examined. In addition, the cells in roots are usually wider, longer, thinner-walled and less lignified, heartwood and tyloses are rare and pith is absent (Fayle 1968, Cutler 1976). As has been demonstrated these differences combine to give the roots of some species an entirely different appearance from their aerial xylem.

In conclusion, with a few exceptions, such as beech, roots are less likely to be of value in dendrochronological studies than trunks and branches.

Acknowledgements

I would like to thank Dr. D.F. Cutler and Dr. P. Rudall for their constructive criticism of the manuscript. Mr. M. Svanderlik and Mr. T. Harwood processed the photographs.

REFERENCES

Baillie, M.G.L. 1982. <u>Tree-Ring Dating and Archaeology.</u> London and Canberra: Croom Helm.

Cutler, D.F. 1976. Variation in Root Wood Anatomy. <u>Leiden Botanical Series, No. 3</u>, pp. 143-156. Leiden University Press.

Fayle, D.C.F. 1968. <u>Radial Growth in Tree Roots: Distribution, Timing, Anatomy.</u> Techn. Rep. No. 9. Faculty of Forestry, University of Toronto.

Fletcher, J.M. 1974. Annual Rings in Modern and Medieval Times. In: <u>The British Oak</u> (eds. Morris, M.G. and Perring, F.H.). Published for The Botanical Society of the British Isles by E.W. Classey Ltd., Farringdon, Berks. pp. 80-97.

Fletcher, J.M. 1975. Relation of Abnormal Earlywood in Oaks to Dendrochronology and Climatology. <u>Nature</u> 254, pp. 506-507.

Gasson, P. 1984. <u>Recognition of Characteristic Features of Secondary Xylem of Selected Hardwoods.</u> PhD. Thesis, University of London (Imperial College).

Knowlson, H. 1939. A Long Term Experiment on the Radial Growth of the Oak. <u>The Naturalist</u> (April 1), pp. 93-99.

Kramer, P.J. and Kozlowski, T.T. 1979. <u>Physiology of Woody Plants.</u> New York, San Francisco, London: Academic Press.

Lebedenko, L.A. 1962. Comparative Anatomical Analysis of the Mature Wood of Roots and Stems of some Woody Plants. <u>Trudy Instituta lesa Akademiya nauk SSSR</u> 51, pp. 124-134 (Translation from Russian by National Lending Library for Science and Technology, England. RTS 2194).

Lohr, E. 1969. Jahresringverlust bei Laubbaumen mit ringporigen Holz. <u>Allgemeine Forst-und Jagdzeitung</u> 140, pp. 18-22 (Forestry Abstracts 30/4599).

Longman, K.A. and Coutts, M.P. 1974. Physiology of the Oak Tree. In: <u>The British Oak</u> (eds. Morris, M.G. and Perring, F.H.) Published for the Botanical Society of the British Isles by E.W. Classey Ltd., Farringdon, Berks.

Schweingruber, F.H. 1978. <u>Microscopic Wood Anatomy. Structural Variability of Stems and Twigs in Recent and Subfossil Woods from Central Europe.</u> Swiss Federal Institute of Forestry Research.

Plate 1. Pedunculate Oak. Transverse sections of secondary xylem. All x 24.5. 1. Trunk, Wistman's Wood, Dartmoor. Cambial age over 100 years. 2. Lateral root from a mature tree, Middlesex. Cambial age less than 10 years. 3. Lateral root from an immature tree about 15 years old, Silwood Park, Berkshire. Cambial age less than 10 years. 4. Tap root from an immature tree about 10 years old, Silwood Park, Berkshire.

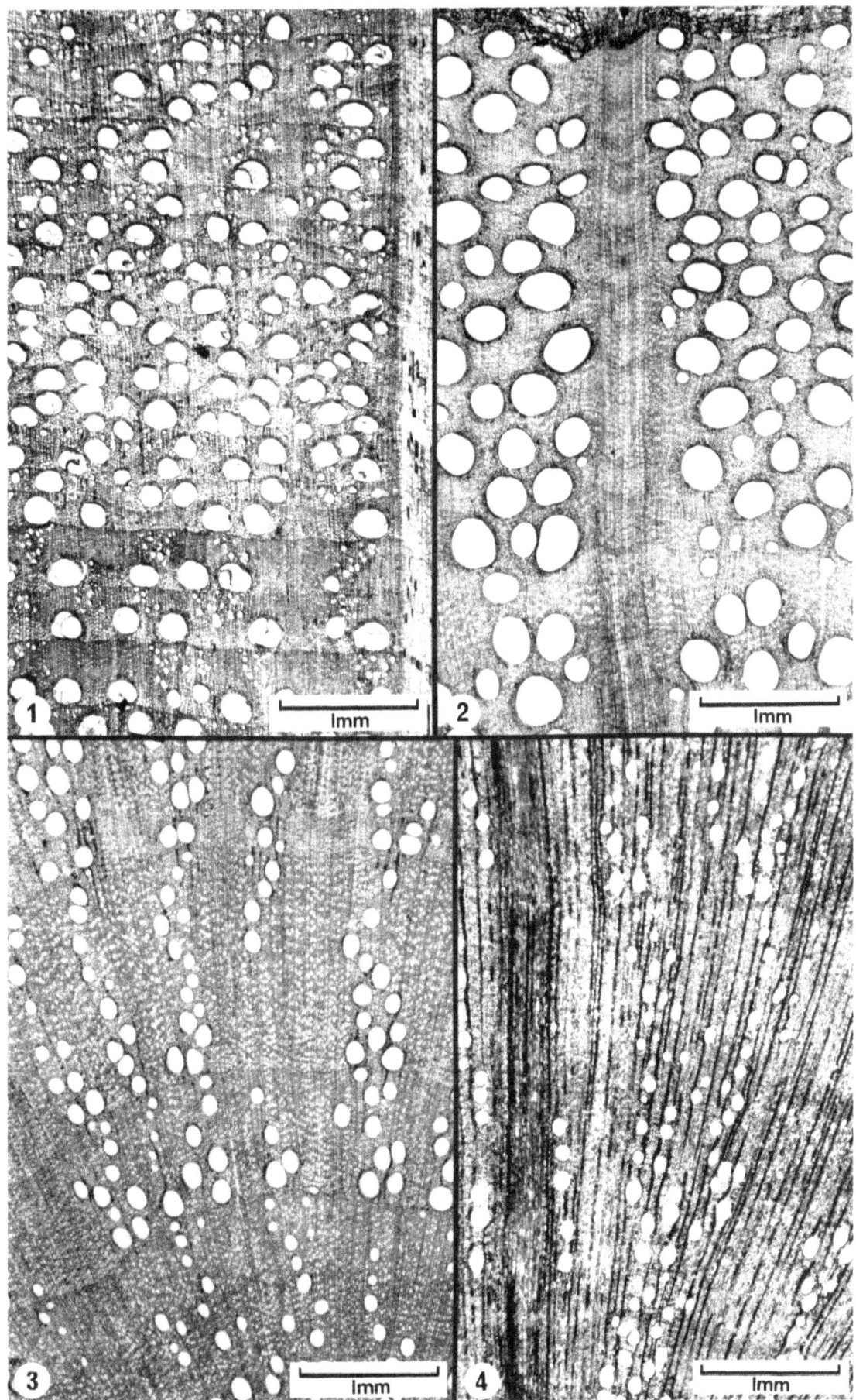

Plate 2. European Beech, (1-3) and Pedunculate Oak (4-5). Transverse sections of secondary xylem. 1, 2, 4 & 5 show a growth ring boundary (x 240), 3 shows two growth rings (x 24.5). 1. Trunk. Cambial age over 100 years. 2. Lateral root, vertically orientated. Cambial age less than 10 years. 3. Trunk at lower magnification showing 1979 and 1980 growth rings. 1-3 from the same beech tree at Kew Gardens, Surrey. 4. Trunk, locality unknown. Cambial age over 100 years. 5. Lateral root, Silwood Park, Berkshire. Cambial age less than 10 years.

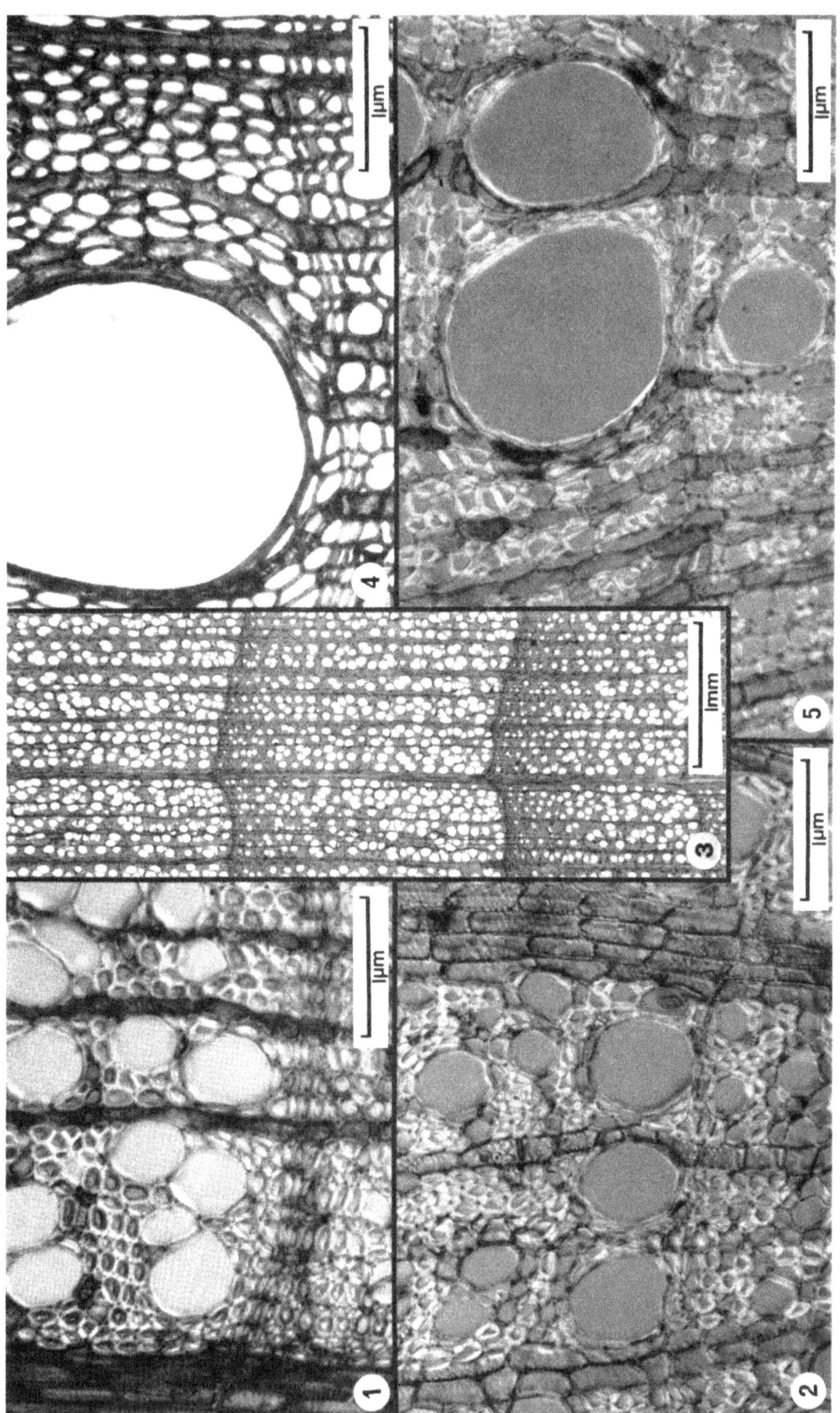

Plate 3. European Ash. Transverse sections of secondary xylem. All x 79. 1. Trunk, site unknown. 2. Branch, Kew Gardens, Surrey. Cambial age 4-10 years old. 3 & 4. Lateral roots, Hendon, Middlesex, both less than 20 years old. It is not known whether they come from the same tree.

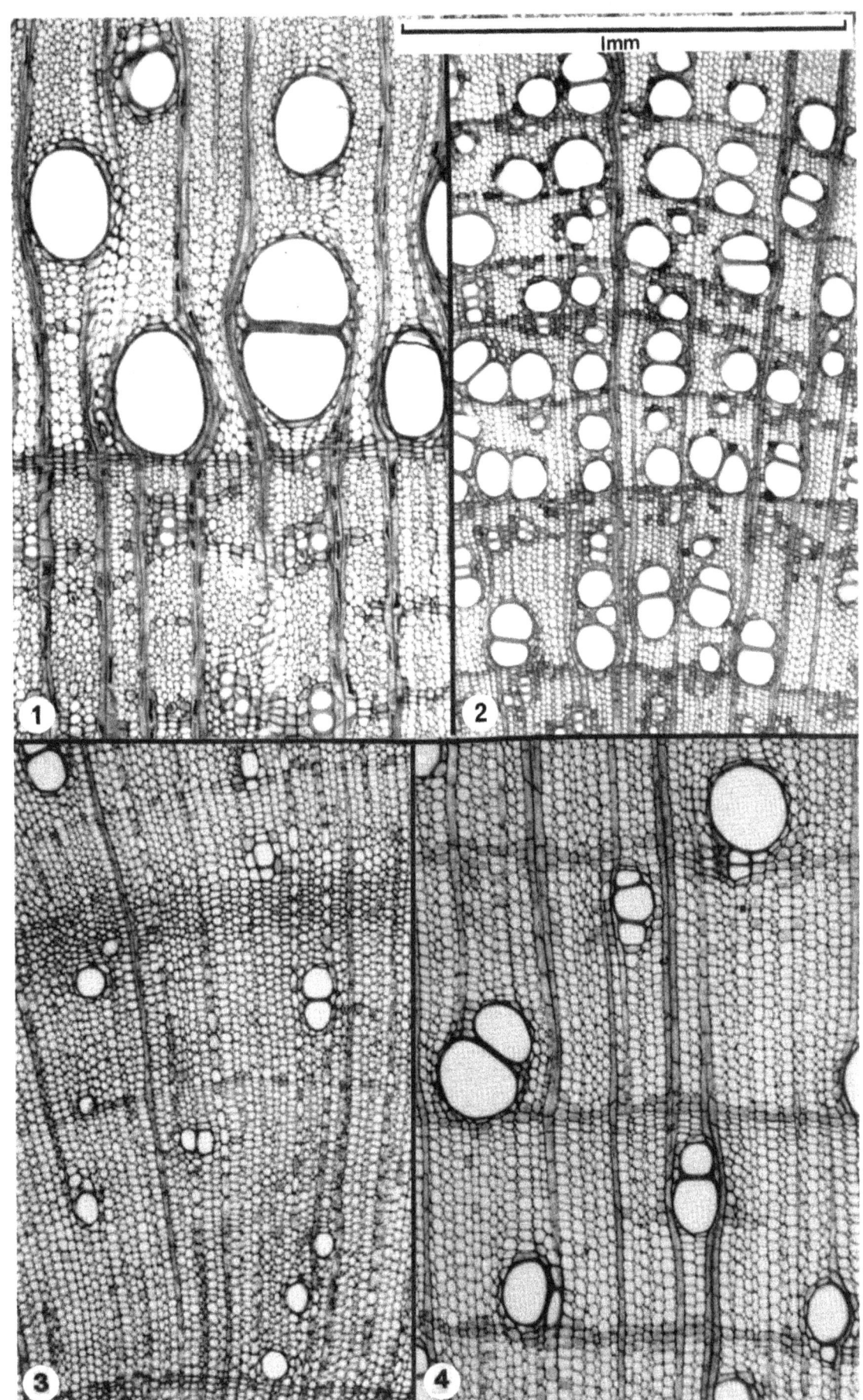

Plate 4. European Ash. Transverse sections of root secondary xylem. All x 24.5. 1. Hendon, Middlesex. 2. Kew Gardens, Surrey. 3. Abinger, Surrey, from a young plant. 4. Site unknown.

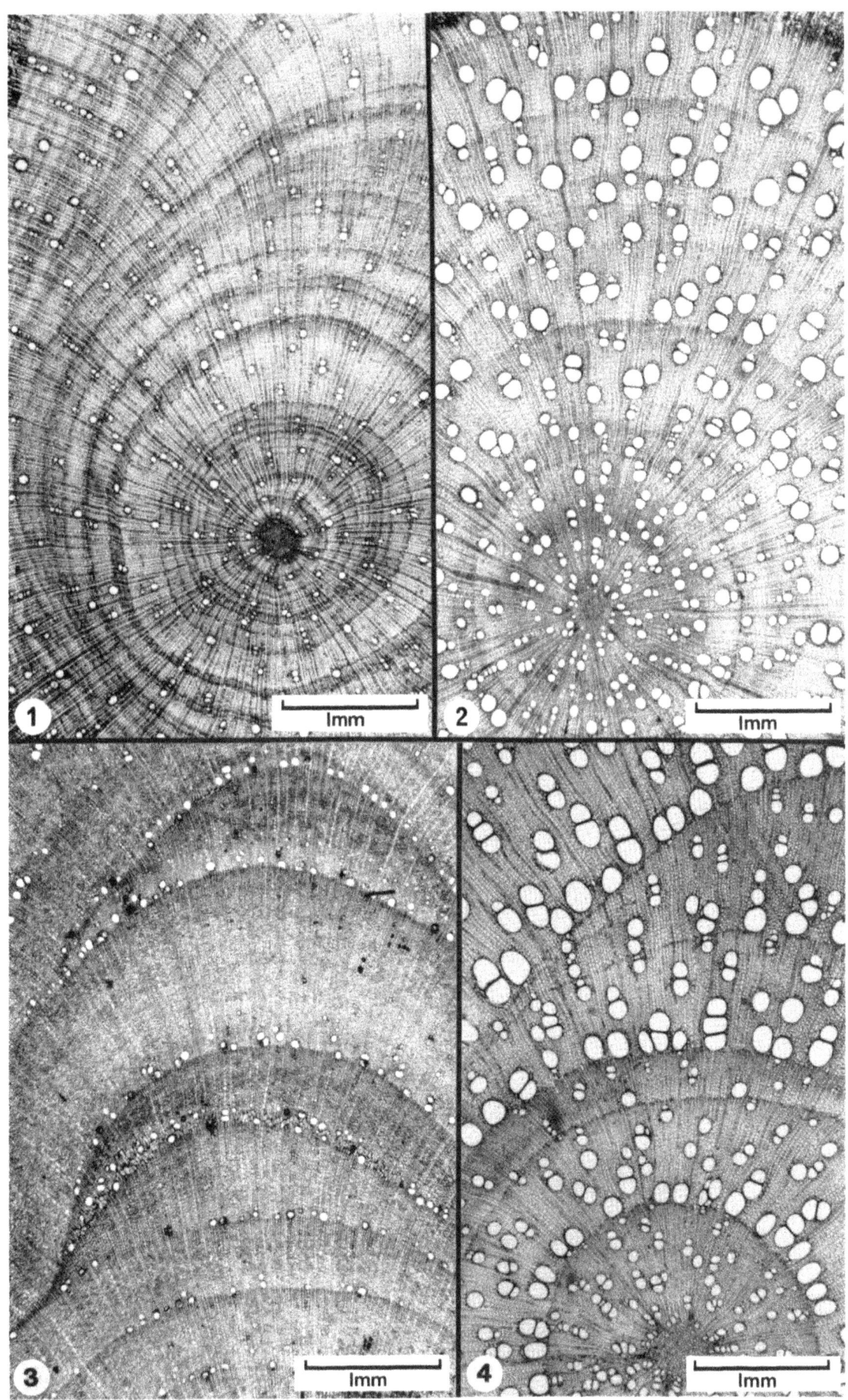

Plate 5. English Elm. Transverse sections of secondary xylem. All x 24.5. 1. Trunk, site unknown, Princes Risborough collection PR385A. 2. Lateral root, Surrey. 3. Lateral root, Essex.

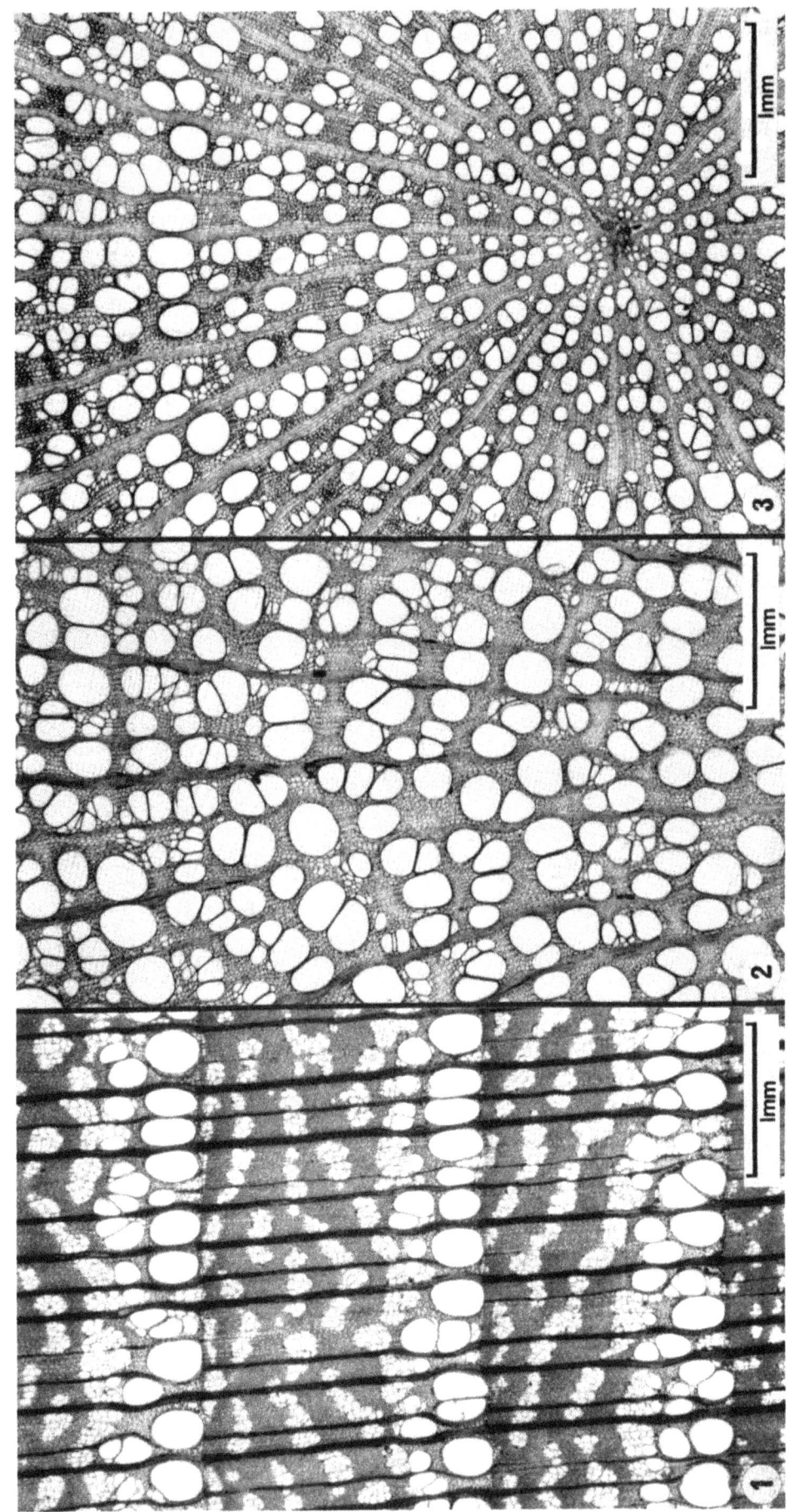

Plate 6. English Elm. Transverse sections of secondary xylem showing growth ring boundaries. All x 79. 1. Trunk, site unknown, Princes Risborough collection, PR385A. 2. Lateral root, Surrey. 3. Lateral root, Essex.

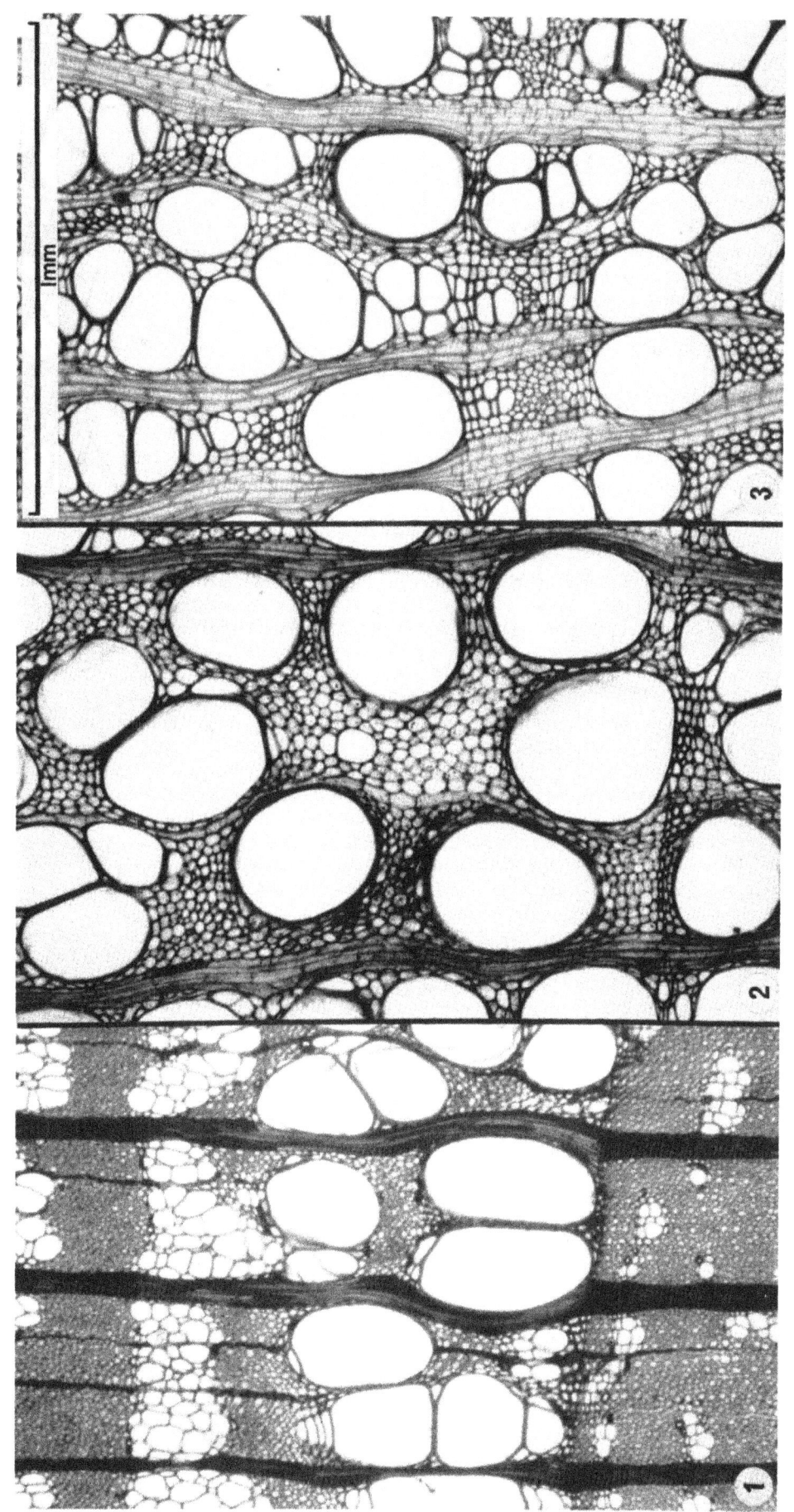

The Palaeoclimatic Significance of Growth Rings in Late Jurassic/
Early Cretaceous Fossil Wood from Southern England

J.E. Francis

Department of Geology & Geophysics
University of Adelaide
Box 498 GPO
Adelaide
South Australia 5001

ABSTRACT

Fossil conifer wood is present as in situ fossil forests in the Upper Jurassic Purbeck Formation in Dorset and in drifted plant beds in the Lower Cretaceous Wealden Group on the Isle of Wight. Analysis of the growth rings in the wood assemblages highlights a climatic change from a semi-arid evaporitic environment in the Purbeck to a humid temperate Wealden climate.

The Purbeck growth rings are narrow and highly variable in width in response to a strongly seasonal climate in which the erratic availability of water markedly affected tree growth. A Mediterranean-type climate, characterised by a seasonal alternation of warm, wet winters and hot, dry summers, is suggested from both tree-ring and sedimentary evidence.

While still reflecting seasonal growth the Wealden rings are slightly wider and more uniform suggesting that the water supply was then less limiting to the growth of the conifers. Other palaeobotanical and sedimentary features also reflect this increase in rainfall. The possible factors causing this climatic change are discussed.

INTRODUCTION

Fossil wood constitutes an important part of many fossil floras. The growth rings in the wood are an important source of palaeoclimatic information as they are one of the few means of recording contemporaneous terrestrial climates and biological productivity. The resolution of some climatic parameters such as annual cyclicity and seasonality can be much more refined than that provided by interpretation of climate-sensitive sediments such as coals and evaporites.

While the interpretation of growth rings in fossil wood has provided important information about changing geological climates on a broad scale (Chaloner and Creber 1973; Creber and Chaloner 1984a, 1984b) the study of individual assemblages of fossil wood ('fossil forests') within the context of their sedimentary environments can produce detailed palaeoclimatic information e.g. the Cretaceous forests of Antarctica (Jefferson 1982) and the Jurassic forests of southern England (Francis 1984). In this paper growth ring analyses of two fossil wood assemblages, the late Jurassic Purbeck fossil forests of Dorset and the early Cretaceous Wealden wood deposits on the Isle of Wight, are used in association with sedimentary evidence to evaluate the changing palaeoclimate at these times.

Geological Setting

i: The Lower Purbeck Fossil Forest

During the late Jurassic, approximately 145 million years ago (Harland et al. 1982), much of southern England was covered by a shallow, highly saline lagoon in which evaporites were deposited (Howitt 1964; West 1979). The Dorset area was situated across the intertidal zone at the margins of the lagoon. Mounds and mats of algal sediment covered this area, merging landward into rendzina-like soils (Francis 1986) which supported gymnosperm forests (Francis 1983). The lagoon water periodically flooded the forests covering the tree stumps with algal-bound sediment.

These sediments are now preserved in the basal part of the Purbeck Formation in Dorset (Figure 1) as a sequence of algal stromatolitic limestones and lagoonal pelletoid siltstones (the 'Caps'), interbedded with carbonaceous clays (the 'Dirt Beds) which represent former forest soils (Arkell 1947; West 1975; Francis 1983). The trees are preserved in the palaeosols as silicified branches and tree strumps which are still rooted in their original growth positions. Their compressed foliage is preserved within the palaeosols and adjacent limestones. The fossil remains indicate that the Purbeck forests were composed mainly of one type of conifer with wood and scale-like leaves similar in appearance to members of the modern Cupressaceae (Francis 1983). However, these trees belonged to the Cheirolepidiaceae, an extinct family of conifers which dominated Mesozoic vegetation (Vakhrameev 1970; Alvin 1982). The growth rings of these fossil trees were studied to investigate the anomalous association of forest vegetation and evaporites within the Purbeck environment.

ii: The Wealden 'Pine Rafts'

Following the relatively quiet Purbeck lagoonal sedimentation, coarse sands and gravels were washed into the basin as a result of uplift and erosion in the surrounding highlands (south-west England, Brittany and the London area) (Allen 1981). The Wealden environment of coastal mudplains and extensive river systems developed, with salinities ranging from freshwater to marine (Allen 1981). A more diverse flora of conifers, cycadophytes and ferns colonised the drained 'upland' areas and river flood-plains (Batten 1974; Hughes 1975; Oldham 1976).

Fossil wood is present as a component of the Brook flora (Hughes 1975) in the Wessex Formation (Barremian stage, approximately 120 million years ago) and is found in abundance on the west coast of the Isle of Wight (Figure 1). The sediments here consist of coarse sands and gravels representing meandering river channels, interbedded with thick variegated marls which were originally deposited on the flood-plains (Stewart 1981).

Plant material consisting of logs, leaves, cones and pollen is preserved within 'plant debris beds', which represent former storm or flood channels which became choked with plant debris. Fossil logs are found in these plant debris beds and also at the bases of the channel sandstones. They range in size from small twigs less than 1cm in diameter to large logs over 1m across. Most wood is preserved as compressed, carbonised lignite but

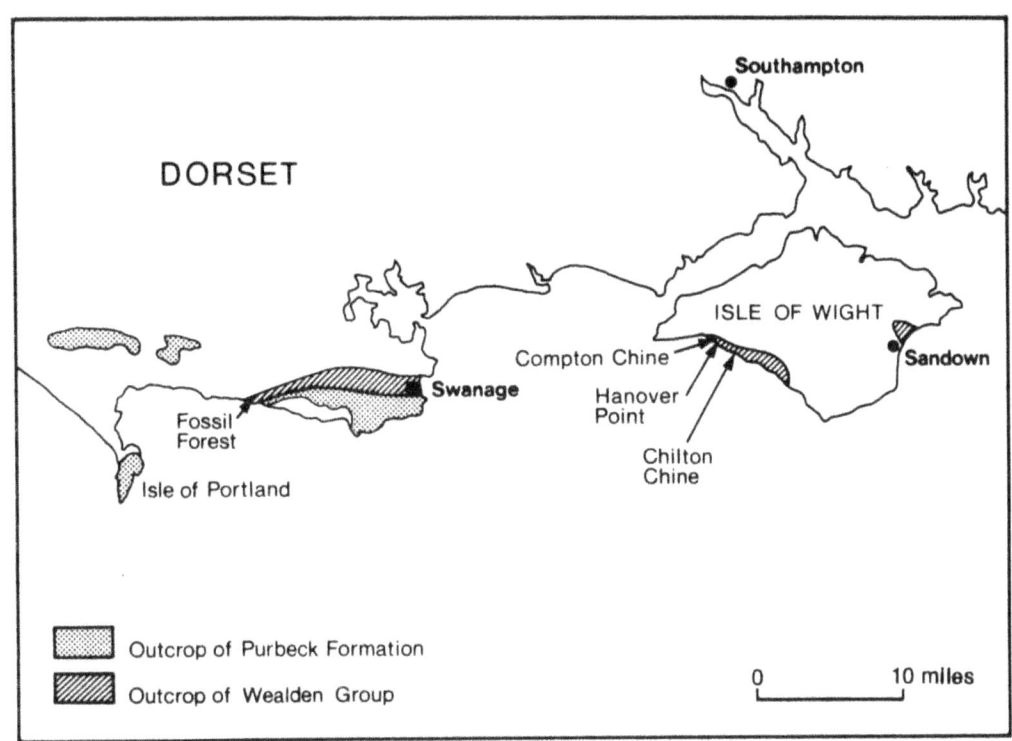

Figure 1. Map showing the outcrops of the Purbeck Formation and the Wealden Group in Dorset and on the Isle of Wight, and some localities mentioned in the text.

Table 1. Summary of results of growth ring analyses of the Purbeck and Wealden fossil wood. For individual specimen data for the Wealden wood see below. For Purbeck tree-ring data see Francis 1984, p.294, Table 2.

Sample	No. of ring series	Mean ring width (mm)	Range of ring widths (mm)	Mean Sensitivity	Range of MS values
PURBECK	20	1.13	0.52-2.28	0.527	0.290-0.778
WEALDEN	9	2.94	0.41-8.90	0.345	0.246-0.443

Wealden tree-ring results:

Sample no.& locality	No. of rings	Mean ring width (mm)	Max. ring width (mm)	Min. ring width (mm)	Mean Sensitivity
W 151 Compton Chine	40	2.27	4.5	0.6	0.362
W 127 Compton Chine	12	2.17	3.9	1.4	0.380
W 150 Compton Chine	8	4.89	6.9	3.4	0.298
W 148 Compton Chine	11	2.44	5.5	2.0	0.259
W 123 Chilton Chine	8	6.96	8.9	4.2	0.285
W 143 Hanover Point	35	2.69	6.1	1.1	0.246
W 101 Compton Chine	37	1.48	3.3	0.3	0.397
W 100 Compton Chine	23	1.54	2.8	0.6	0.439
W 102 Compton Chine	22	2.00	4.3	0.8	0.443

some has been permineralised with silica or calcite and retains its cellular detail. The most well-known of these wood deposits is the 'Pine Raft' at Hanover Point (Figure 1) where large calcified trunks are preserved as a log jam, exposed on the foreshore at low tide. Although the trees were transported away from their growth sites the excellent preservational state of the plant material (Oldham 1976; Alvin et al 1981) and the known palaeogeography (Allen 1981; Stewart 1981) implies that the wood had only drifted a short distance before being buried.

The fossil wood from the Wessex Formation belongs to several form-genera of fossil conifers (Alvin et al 1981). One Wealden conifer Pseudofrenelopsis was reconstructed by Alvin (1983) as a tall forest tree with scale-like foliage, similar to the Purbeck tree (Francis 1983, fig. 3) and also a member of the Cheirolepidiaceae.

GROWTH RING ANALYSIS

The growth rings were measured from polished slabs of fossil wood in transverse section using a binocular microscope, from acetate peels of acid-etched surfaces and from petrographic thin-sections. The rings were measured along a radial line to obtain as long a series as preservation permitted, although readings often had to be made along adjacent radii to avoid patches of poor cellular preservation. Often only a portion of the original diameter of the log was present so the radii of curvature of the rings were noted to estimate whether the wood specimen came from a large trunk or small branch.

Some of the standard statistical parameters used to describe modern ring characteristics (Fritts 1976; Creber and Chaloner 1984b) were used for this study. However, some adaptions had to be made for the additional problems associated with fossil wood such as the use of only fragments of wood with a limited number of rings, an unknown position within the tree and very limited knowledge of local site factors which had affected growth (Creber and Chaloner, this volume). The presence or absence of rings was noted and the average ring width calculated as an indication of seasonal growth rates. The variability of the ring widths was the most informative parameter, illustrating the sensitivity of tree growth to climatic influence. This was expressed in terms of Mean Sensitivity (MS) and calculated using the formula:

where X is the ring width, n is the number of rings and t is the year number of the ring (Fritts 1976). The MS values range from 0 where there is no variation from year to year, to a theoretical maximum of 2 representative of the greatest variation. An arbitrary value of 0.3 is taken to separate "complacent" trees which have had little climatic influence (MS 0.3) from those which were "sensitive" to climatic variation (MS 0.3) (Fritts 1976). The Annual Sensitivity (AS), the difference in width between a pair of consecutive rings divided by their average width (Creber 1977), was also used to determine the nature of the annual variation.

For selected rings with good cellular preservation individual cell diameters along radial files were measured in order to classify the rings according to the scheme proposed for fossil wood by Creber and Chaloner (1984b, p. 371). The cumulative algebraic sum of each cell's deviation from the mean of the radial cell diameters was calculated and the graphs in Figure 2 drawn.

Results

Ring series consisting of 8 to 105 rings were obtained, although continuous series of 20 - 30 rings were most frequent. The results are summarised in Table 1. Conspicuous growth rings were present in both the Purbeck and Wealden wood, and in all trees the rings consisted of a relatively wide zone of large, thin-walled earlywood cells, followed by a very narrow zone of only 4 - 5 thick-walled latewood cells (Figure 2). This pattern is characteristic of the ring types D and E illustrated by Creber and Chaloner (1984b, p. 373), and indicates a relatively uniform growing season followed by a marked terminal event representing a cessation or retardation of cambial activity.

The Purbeck growth rings were much narrower (average 1.13mm) than those of the Wealden (average 2.94mm); the largest Purbeck ring (2.28mm) being considerably smaller than the maximum size of the Wealden rings (8.90mm). However, the most striking feature of the Purbeck rings is their great variation in width from year to year (Figure 3). This is reflected in the high Mean Sensitivity values, which often exceeded 0.7. These trees were clearly highly sensitive to climatic stress and irregular growth conditions. Such high values are characteristic of trees from semi-arid forests (Schulman 1956) and particularly from the lower borders of semi-arid forests where water availability is a limiting factor to growth (Fritts et al. 1965). The erratic availability of water is also the most obvious cause of irregular tree growth in the evaporitic Purbeck environment. Consistent with this is the frequent presence of false, or intra-annual rings. These are bands of small, dense cells within the zone of large earlywood cells (Francis 1984, Figure 5d), very similar to drought rings (Glerum 1970).

The Wealden trees also have a 'sensitive' response to limiting growth factors (the average figure for Mean Sensitivity is 0.345). However, Figure 4 illustrates that a high proportion of trees show 'complacent' ring sequences as a result of more uniform growth. False rings are less frequent in the Wealden wood. The histograms of Annual Sensitivities (Figure 5) reflect the much greater degree of annual variation in growth of the Purbeck trees than the Wealden. In the Wealden trees 52.5% of the Annual Sensitivity values are 'sensitive', compared to 63.9% in the Purbeck trees, again reflecting the more uniform growth in the Wealden trees.

Interpretation of the growth rings

The presence of growth rings in these Mesozoic woods is evidence that the climate in southern England at that time had well-defined seasons, and was unlikely to have been a seasonless, tropical environment. The narrowness of the Purbeck rings implies that these trees grew rather slowly whereas the Wealden trees had more rapid growth rates. Both sets of trees were influenced by rather fluctuating climates, the most important factor probably being the availability of water which suggests variable rainfall. The Purbeck trees appear to have suffered more from water stress, not only

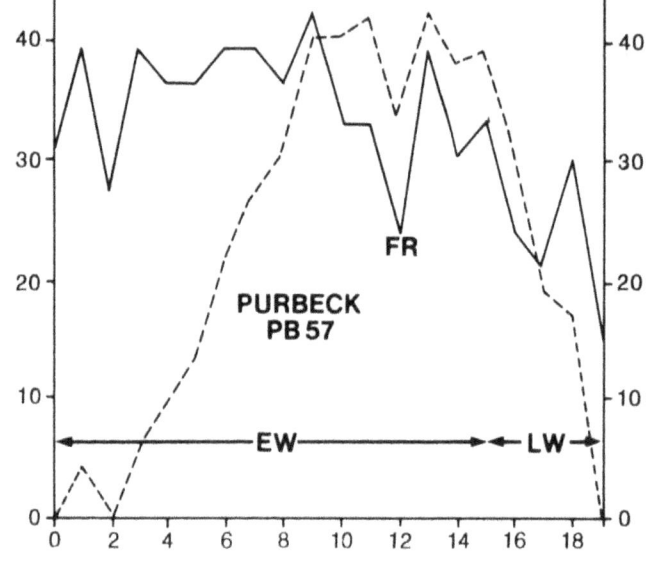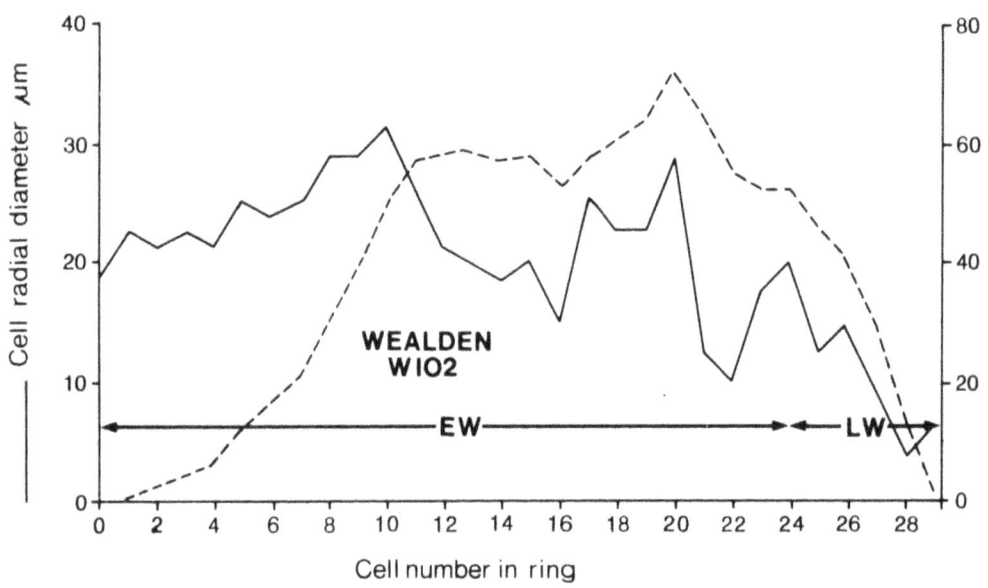

Figure 2. Graphs showing the change in cell radial diameter across selected growth rings in Purbeck and Wealden wood. The earlywood (EW)/latewood (LW) boundary is chosen as the point at which the cumulative sum finally returns to zero. A marked decrease in cell diameter in the earlywood zone of the Purbeck ring marks the position of a false ring (FR). These rings are characteristic of the types D and E of Creber and Chaloner 1984b.

 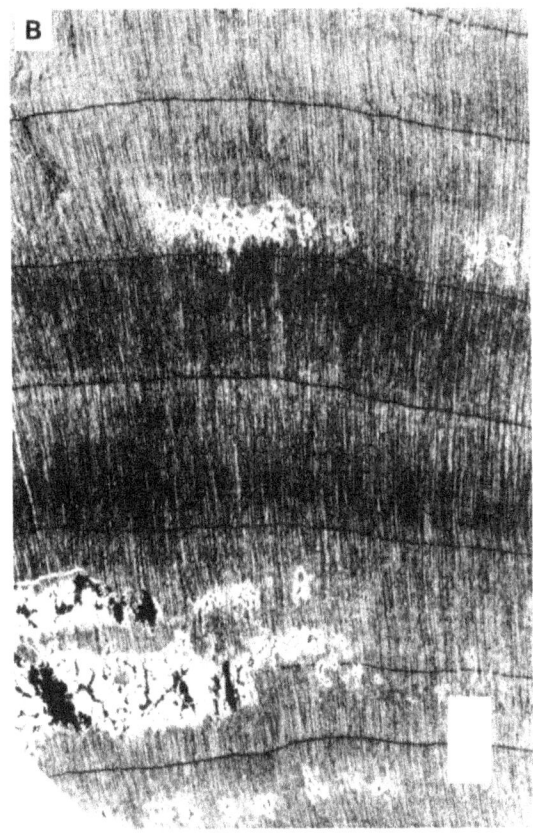

Figure 3. (a) Growth rings in Purbeck fossil wood. PB 24, Portland. Thin-section of silicified wood. Note the very variable widths of the rings from year to year.
(b) Growth rings in Wealden fossil wood. W123, Brook flora, Isle of Wight. Thin-section of silicified wood. These rings are much wider and more uniform in width. Each ring consists of a broad zone of earlywood cells terminated by a narrow ring of only 4-5 latewood cells. Scale bars represent 5 mm.

due to lack of water at the end of the growing season but also from intermittent drought during the growing season itself, as implied by the false rings.

Comparison with the growth patterns of living trees of known climatic influence has shown that trees from markedly seasonal Mediterranean-type climatic zones (with warm, wet winters and hot, dry summers) have ring characteristics most similar to the Purbeck ring series (Francis 1984). In particular they show the same slow growth rates and the same kind of variability of annual growth (and thus high Mean Sensitivities), and these characteristics can be related to the erratic occurrence and quantity of rainfall within the wet season. The Wealden growth rings continued to exhibit this irregular growth but to a lesser degree. The greater width of the rings and the fewer false rings suggests that at this time seasonality did not have such a severe effect on these trees, possibly because the rainfall was higher.

SEDIMENTARY EVIDENCE OF SEASONAL CLIMATES IN THE PURBECK AND WEALDEN ENVIRONMENTS

Purbeck

Support for these conclusions is provided by several aspects of the sediments and fossils in the basal part of the Purbeck Formation which also suggest the influence of a seasonal climate.

Firstly, a rather anomalous collection of silicified wood, seeds and charophytes (Barker et al. 1975) of freshwater affinities, along with nodules of silicified evaporite pseudomorphs of hypersaline origin can be found in the western part of the Purbeck outcrop (West 1975). Similar assemblages can be found today in coastal lagoons in Mediterranean climatic zones of southern Australia (Burne et al. 1980), and are able to co-exist because the seasonal alternation of rainfall and drought changes the salinity of the lagoon water from fresh in winter to hypersaline in the summer. This Australian environment is thus considered analogous to that of the Purbeck (West 1975; Burne et al. 1980).

Secondly, the quartz that is present in the silicified Purbeck wood is a type of chalcedony (quartzine) often associated with highly alkaline environments. It is frequently used as a palaeoclimatic indicator of semi-arid climates (Folk and Pittman 1971). The source of the silica seems most likely to have been as an inorganic precipitate influenced by alternating changes in the lagoon water pH; a modern model related to seasonality changes for this mechanism can again be found in the Australian lagoons (Peterson and Von Der Borch 1965).

Thirdly, at the base of the rendzina-like fossil soil of the Great Dirt Bed there are irregularly laminated nodules and crusts of fine-grained carbonate, commonly known as calcrete (Francis, 1986). A seasonal semi-arid climate with periods of rainfall followed by periods of intense evaporation is always necessary for the formation of this pedogenic calcrete (Harrison 1977).

Finally, within the Lower Dirt Bed in east Dorset, which here represents a marginal lagoonal deposit, assemblages of small conchostracan branchiopod crustacea (fossil clam shrimps) can be found on lamina surfaces (Francis

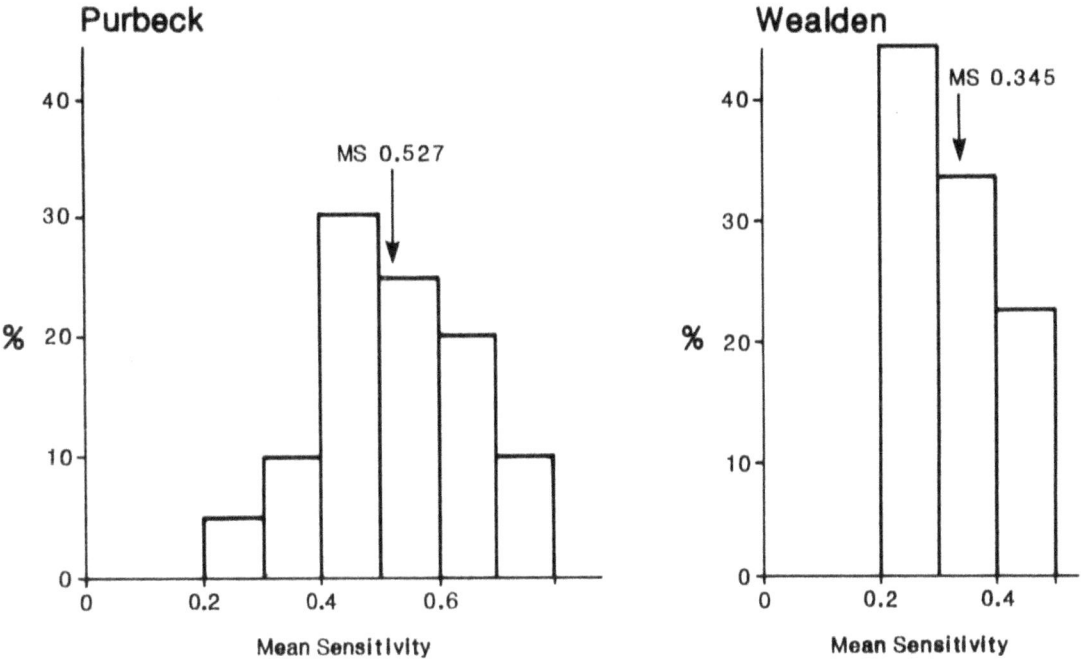

Figure 4. Histogram of Mean Sensitivity values for the Purbeck and Wealden fossil wood. Over 95% of the Purbeck trees have 'sensitive' mean sensitivity values (MS > 0.3) compared to only 56% of the Wealden trees.

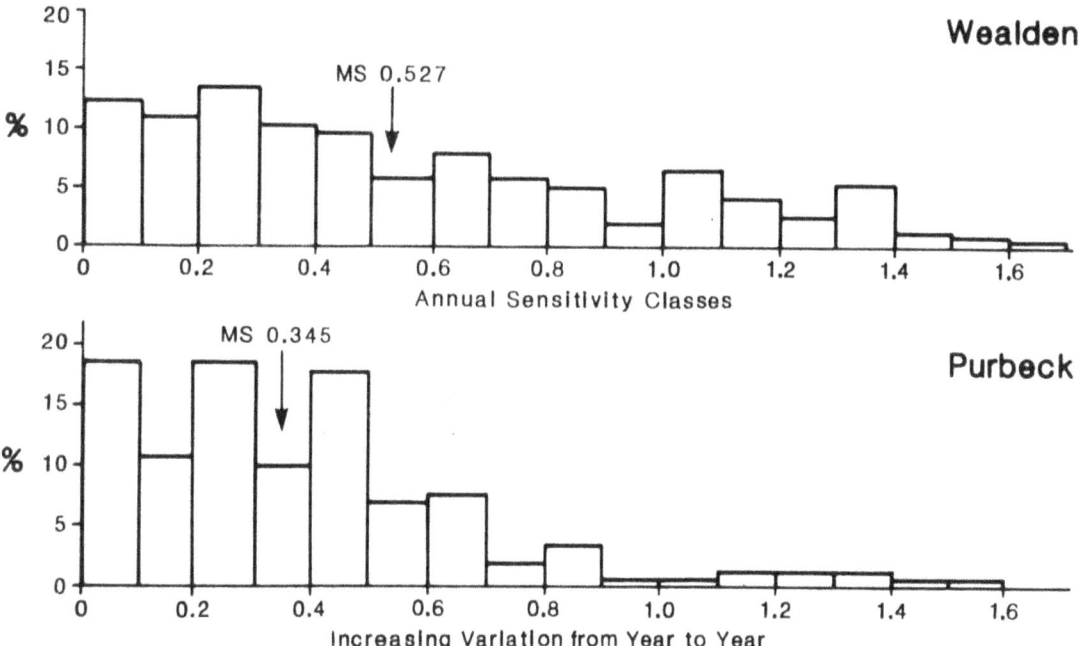

Figure 5. Histograms of Annual Sensitivity values for the Purbeck and Wealden trees. The range in annual variation is much greater in the Purbeck trees than in the Wealden, in response to a more irregular climate. (MS = mean sensitivity).

1984). Their modern representatives live in small brackish or freshwater pools which often desiccate during the dry season, leaving 'death assemblages' on the sediment (Tasch and Zimmerman 1960). The number of concentric moult lines on the carapaces can be used to determine the life span of the crustacea (Tasch 1969). By analogy the Purbeck crustacea may have lived for about 5 months, which may also represent the duration of the wet season.

Wealden

Alternative evidence exists to support the notion of a wetter Wealden period. The increased proportion of river, storm and flood sediments indicates that river systems became a dominant factor of the Wealden landscape, implying that uplift and erosion of the adjacent highlands and a subsequent increase in rainfall had occurred since the Purbeck times. Further, the change to a more humid climate was envisaged by Batten (1974) and Hughes (1975) to account for a more diverse flora with a higher proportion of ferns than the Purbeck. The marked decrease in the amount of Classopollis pollen (yielded by members of the Cheirolepidiaceae conifer family), which is abundant in the Purbeck sediments and associated with semi-arid, evaporitic Mesozoic environments worldwide (Vakhrameev 1970; Srivastava 1976), also signifies a wetter climate (Batten 1982).

However the sediments indicate that the rainfall was not constant throughout the year. Allen (1981, p. 379) envisaged a "warm and periodically wet" climate to account for the abundance of storm and flood sediments induced by periods of heavy rainfall and to maintain lakes inhabited by freshwater invertebrates. The fluctuating discharge of the rivers was attributed to an irregular rainfall pattern. Although intermittently very wet, desiccation features such as mudcracks within channel base sediments are evidence that the water levels periodically became very low during the dry season (Stewart 1983). During these times the vegetation became sufficiently dry to fuel forest fires, so producing the charcoal which is common in Wealden sediments. Most of the charcoal is formed of conifer wood (Alvin et al. 1981) or burnt fern stems (Harris 1981). The ferns probably grew on the river flood-plains but were rapidly buried after the fire, ensuring their preservation. Harris (1981) considered that the climate probably consisted of a long dry season followed by a short but stormy wet season. Drought conditions were not of sufficient intensity or duration, however, to allow the precipitation of evaporites which are notably absent in the Wealden sediments. Oxygen-isotope analyses on mollusc shells indicate that the Wealden temperatures were comparable to those of modern temperate/tropical climates (Allen et al. 1973).

DISCUSSION

The sediments and fossils record a change from a semi-arid, evaporitic environment during the Purbeck to a more humid temperate Wealden setting in the early Cretaceous. The floras also reflect this transition, the arid Purbeck conifer forests which grew in well-drained soils (Francis 1984) being replaced by the fern and lycopod type vegetation of the Wealden flood-plains (Batten 1974). Further details of this change were obtained by Sladen (1983) and Sladen and Batten (1984) from palynofacies data and clay-mineral suites. The purbeck beds contain a high proportion of illite and mixed-layer clay minerals with only a little kaolinite, a suite found today in alkaline, poorly-leached soils of semi-arid areas. Throughout the early Cretaceous kaolinite increases in adundance, reflecting the increasing humidity and leaching in more acid conditions. By comparison with the climatic parameters governing the distribution of modern clay mineral suites Sladen and Batten (1984) proposed that the Purbeck climate was arid/semi-arid with mean temperatures of 20–25°C and rainfall less than 500mm per year. Likewise the Wealden (Barremian) climate was likely to have been warm humid-temperate with similar temperatures but rainfall probably over 1000mm per year.

The growth patterns in the fossil wood also illustrate that the trees reflected this climatic transition too. The Purbeck growth rings are narrow and variable in width in response to a rather harsh semi-arid environment in which the erratic availability of water adversely affected growth. The rainfall was estimated to be about 400mm per year by Francis (1983) based on sedimentary analysis and by comparison with a very similar living conifer forest in a lagoonal setting on Rottnest Island, west Australia. The fossil wood in the Wessex Formation, however, has notably wider and more uniform growth rings suggesting that larger quantities of water were available, particularly in the seasons important for tree growth. The rainfall estimate of 1000mm by Sladen and Batten (1984) seems appropriate for growth of the Wealden trees also.

Although the sedimentary evidence does imply an irregular distribution of rainfall in both Purbeck and Wealden times, the only means of assessing the exact nature of the seasonality is from the growth patterns in the fossil trees. The characteristics of individual rings highlight an annual pattern consisting of a favourable season of growth followed by a period in which the tree became almost dormant (Figure 2). For the Purbeck trees this has been interpreted in terms of a seasonal climate with warm, wet winters followed by hot, arid summers, that is, a Mediterranean-type climate (Francis 1984). The period of drought was quite severe in this case to allow gypsiferous evaporites to form (West 1979). Even the winter season was somewhat harsh with intermittent dry spells causing drought rings to form in the trees. The growth ring patterns in the Wealden trees, though less sensitive, continue to show the same kind of annual pattern and suggest that the same kind of seasonal climate continued into the early Cretaceous, though probably somewhat ameliorated by higher rainfall which allowed the trees to grow more steadily.

This climatic pattern probably reflects local conditions over southern England. Globally, the Mesozoic climate had higher ambient temperatures than today with an expanded tropical belt and temperate zones extending to the poles (Frakes, 1979; Hallam 1984; 1985). The continents were then joined to form a large land mass (Pangaea) which was subject to a much more continental climate with extremes of winter and summer temperatures (Frakes 1979; Barron and Washington 1982). Palaeoclimatic simulation models of the Cretaceous climate (Lloyd 1982; Barron and Washington 1982; Parrish et al. 1982) predicted alternating high and low pressure centres over the continents and oceans, resulting in a strongly seasonal climate over the continental margins. The growth rings in the fossil wood support this idea. Britain was then situated near the western borders of the Tethyan ocean, which formed between the Asian continent to the north and the Gondwanan land mass to the south (Hallam 1984).

The change from a semi-arid to a more humid climate in southern England was unlikely to have been due to a latitudinal shift into a more northerly climatic zone since the palaeolatitudinal position of this area for both Purbeck and Wealden times remains constant at about 36°N (Smith et al. 1981). An alternative explanation is local uplift of adjacent highlands (Sladen and Batten 1984, Allen 1981), coupled with an increasing maritime influence due to the westerly extension of the Tethys seaway and the initial opening of the North Atlantic (Hallam 1984).

SUMMARY

During the late Jurassic and early Cretaceous the vegetation in southern England consisted of conifer forests which are now preserved as the Purbeck in situ fossil forests and the Wealden drifted plant beds. The growth rings in the fossil wood show that the trees grew in a seasonal, temperate environment. The tree-rings in the Purbeck wood are narrow and variable in width from year to year, reflecting growth in a markedly seasonal, semi-arid climate where the erratic availability of water strongly influenced tree growth. The climate seems to have been of Mediterranean-type with warm, wet winters suitable for the growth of the trees followed by hot, dry summers when the evaporites formed. The rainfall is estimated to have been about 400 - 500 mm per year, based on both floral and sedimentary evidence.

Growth rings in the wood and analysis of the associated sedimentary rocks indicate that the same pattern of seasonality continued into Wealden times. However the greater widths of the rings and the slightly lower (though still 'sensitive') Mean Sensitivity values suggests that the rainfall was significantly higher than during the Purbeck. The proportion of fluvially-influenced and kaolinite-rich sediments is evidence in support of this. The annual rainfall was estimated to be over 1000 mm.

This seasonality corresponds with palaeoclimatic simulation models which predict the presence of seasonal climates over continental margins in the Mesozoic. The increase in rainfall from Purbeck to Wealden times has been related to the increased elevation of the surrounding highlands and the increasing maritime influence related to the opening of the Tethyan seaway and the North Atlantic ocean.

Acknowledgements

I am grateful for research facilities provided at Southampton University and Bedford College (University of London) during the tenure of NERC research grants, and the British Antarctic Survey. I am indebted to Dr. G.T. Creber for valuable advice and discussion.

REFERENCES

Allen, P. 1981. Pursuit of Wealden models. The Journal of the Geological Society 138, pp. 375-405.

Allen, P., Keith, M.L., Tan, F.C. and Deines, P. 1973. Isotopic ratios and Wealden environments. Palaeontology 16, pp. 607-621.

Alvin, K.L. 1982. Cheirolepidiaceae: biology, structure and palaeoecology. Review of Palaeobotany and Palynology 37, pp. 71-98.

Alvin, K.L. 1983. Reconstruction of a Lower Cretaceous conifer. Botanical Journal of the Linnean Society 86, pp. 169-176.

Alvin, K.L., Fraser, C.J. and Spicer, R.A. 1981. Anatomy and palaeoecology of Pseudofrenelopsis and associated conifers in the English Wealden. Palaeontology 24, pp. 759-778.

Arkell, W.J. 1947. The geology of the country around Weymouth, Swanage, Corfe and Lulworth. Memoir of the Geological Survey of Great Britain.

Barker, D., Brown, C.E., Bugg, S.C. and Costin, J. 1975. Ostracods, land plants and charales from the basal Purbeck Beds of Portesham Quarry, Dorset. Palaeontology 18, pp. 419-436.

Barron, E.J. and Washington, W.M. 1982. Cretaceous climate: a comparison of atmospheric simulations with the geologic record. Palaeogeography, Palaeoclimatology, Palaeoecology 40, pp. 103-133.

Batten, D.J. 1974. Wealden palaeoecology from the distribution of fossil plants. Proceedings of the Geologists Association 85, pp. 433-458.

Batten, D.J. 1982. Palynofacies and salinity in the Purbeck and Wealden of southern England. In: Aspects of Micropalaeontology (eds. Banner, F.T. and Lord, A.R.) London: George Allen and Unwin, pp. 279-295.

Burne, R.V., Bauld, J. and DeDecker, P. 1980. Saline lake charophytes and their geological significance. Journal of Sedimentary Petrology 50, pp. 281-293.

Chaloner, W.G. and Creber, G.T. 1973. Growth rings in fossil woods as evidence of past climates. In: Implications of Continental Drift to the Earth Sciences. (eds. Tarling, D.H. and Runcorn, S.K.) London: Academic Press, pp. 425-437.

Creber, G.T. 1977. Tree rings: a natural data-storage system. Biological Reviews of the Cambridge Philosophical Society 52, pp. 349-383.

Creber, G.T. and Chaloner, W.G. 1984a. Climatic indications from growth rings in fossil woods. In: Fossils and Climate. (ed. Brenchley, P.) Chichester: John Wiley and Sons Ltd., pp. 49-74.

Creber, G.T. and Chaloner, W.G. 1984b. Influence of environmental factors on the wood structure of living and fossil trees. The Botanical Review 50, pp. 357-448.

Creber, G.T. and Chaloner, W.G. 1987 (this volume). The contribution of growth ring studies to the reconstruction of past climates. pp. 37-67.

Folk, R.L. and Pittman, J.S. 1971. Length-slow chalcedony: a new testament for vanished evaporites. Journal of Sedimentary Petrology 4, pp. 1045-1058.

Frakes, L.A. 1979. Climates throughout geologic time. Amsterdam: Elsevier.

Francis, J.E. 1983. The dominant conifer of the Jurassic Purbeck Formation, England. Palaeontology 26, pp. 277-294.

Francis, J.E. 1984. The seasonal environment of the Purbeck (Upper Jurassic) fossil forests. Palaeogeography, Palaeoclimatology, Palaeoecology 48, pp. 285-307.

Francis, J.E. (1986). The calcareous paleosols of the basal Purbeck Formation (Upper Jurassic), southern England. In: Paleosols (ed. Wright, V.P.) Princeton: Blackwells and Princeton University Press. pp. 112-138.

Fritts, H.C. 1976. Tree rings and climate. London: Academic Press.

Fritts, H.C., Smith, D.G., Cardis, J.W. and Budelsky, C.A. 1965. Tree-ring characteristics along a vegetation gradient in northern Arizona. Ecology 46, pp. 393-401.

Glerum, C. 1970. Drought ring formation in conifers. Forest Science 16, pp. 246-248.

Hallam, A. 1984. Continental humid and arid zones during the Jurassic and Cretaceous. Palaeogeography, Palaeoclimatology, Palaeoecology 47, pp. 195-223.

Hallam, A. 1985. A review of Mesozoic climates. The Journal of the Geological Society 142, pp. 433-446.

Harland, W.B., Cox, A.V., Llewellyn, P.G., Pickton, C.A.G., Smith, A.G. and Walters, R. 1982. A geologic time scale. Cambridge: Cambridge University Press.

Harris, T.M. 1981. Burnt ferns from the English Wealden. Proceedings of the Geologists Association 92, pp. 47-58.

Harrison, R.S. 1977. Caliche profiles: indicators of near-surface subaerial diagenesis, Barbados, West Indies. Bulletin of Canadian Petroleum Geology 25, pp. 123-173.

Howitt, F. 1964. Stratigraphy and structure of the Purbeck inliers of Sussex, (England). Quarterly Journal of the Geological Society of London. 120, pp. 77-113.

Hughes, N.F. 1975. Plant succession in the English Wealden. Proceedings of the Geologists Association 86, pp. 439-455.

Jefferson, T.H. 1982. Fossil forests from the Lower Cretaceous of Alexander Island, Antarctica. Palaeontology 25, pp. 681-708.

Lloyd, C.R. 1982. The mid-Cretaceous earth: palaeogeography; ocean circulation and temperature; atmospheric circulation. Journal of Geology 90, pp. 393-413.

Oldham, T.C.B. 1976. Flora of the Wealden plant debris beds of England. Palaeontology 19, pp. 437-502.

Parrish, J.T., Ziegler, A.M. and Scotese, C.R. 1982. Rainfall patterns and the distribution of coals and evaporites in the Mesozoic and Cenozoic. Palaeogeography, Palaeoclimatology, Palaeoecology 40, pp. 67-102.

Peterson, M.N.A. and Von der Borch, C.C. 1965. Chert: modern inorganic deposition in a carbonate-precipitating locality. Science 149, pp. 1501-1503.

Schulman, E. 1956. Dendroclimatic changes in semi-arid America. Arizona: University of Arizona Press.

Sladen, C.P. 1983. Trends in Early Cretaceous clay mineralogy in NW Europe. Zitteliana 10, pp. 349-357.

Sladen, C.P. and Batten, D.J. 1984. Source-area environments of Late Jurassic and Early Cretaceous sediments in Southeast England. Proceedings of the Geologists Association 95, pp. 149-163.

Smith, A.G., Hurley, A.M. and Briden, J.C. 1981. Phanerozoic Palaeocontinental World Maps. Cambridge University Press.

Srivastava, S.K. 1976. The fossil pollen genus Classopollis. Lethaia 9, pp. 437-457.

Stewart, D.J. 1981. A field guide to the Wealden Group of the Hastings area and the Isle of Wight. In: Field guides to modern and ancient fluvial systems in Britain and Spain. (ed. Elliot, T.). International Conference on Fluvial Systems. Keele.

Stewart, D.J. 1983. Possible suspended-load channel deposits from the Wealden Group (Lower Cretaceous) of Southern England. Special Publications of the International Association of Sedimentologists 6, pp. 369-384.

Tasch, P. 1969. Branchiopoda. In: Treatise on Invertebrate Palaeontology. Part R Arthropoda (ed. Moore, R.C.).

Tasch, P. and Zimmerman, R. 1960. Fossil and living conchostracan distribution in Kansas-Oklahoma across a 200 million year time gap. Science 133, pp. 584-586.

Vakhrameev, V.A. 1970. Range and palaeoecology of Mesozoic conifers, the Cheirolepidiaceae. Palaeontological Journal 1970/1, pp. 12-25.

West, I.M. 1975. Evaporites and associated sediments of the basal Purbeck Formation (Upper Jurassic) of Dorset. Proceedings of the Geologists Association 86, pp. 205-225.

West, I.M. 1979. Review of evaporite diagenesis in the Purbeck Formation of Southern England. Symposium on Jurassic sedimentation in Western Europe. <u>Association des Sedimentologistes Francais. Publication speciale.</u> 1, pp. 407-415.

The Contribution of Growth Ring Studies to the Reconstruction of
Past Climates

G.T. Creber and W.G. Chaloner

Botany Department
Royal Holloway and Bedford New College
Egham Hill, Egham
Surrey TW20 OEX

ABSTRACT

The fossil remains of the earliest known trees from the Upper Devonian and Carboniferous Periods are either without growth rings or only have them very weakly developed. Since these tree grew largely within the low palaeolatitudes of those Periods, their ring characteristics seem quite consistent with growth in an almost seasonless climate. In more recent geological periods, particularly the Mesozoic, fossil wood with very wide rings is found in very high palaeolatitudes where at the present day tree growth is either very limited or absent altogether. It seems likely that the main climatic difference between the present day and the Mesozoic high latitudes is the low ambient temperature which depresses metabolic processes such as transport and growth rather than photosynthesis. For the latter, the supply of light energy at high latitudes would appear to be adequate to explain the high productivity in the Mesozoic provided there had been higher ambient temperatures.

INTRODUCTION

Fossil wood is found in great abundance with a world-wide distribution. This ubiquity of fossil wood is due to a number of factors: 1. Wood is formed in considerable bulk; 2. It is a relatively resistant material which may survive undecayed on land or in water for some time before fossilization; 3. It appears to have a positive attraction for ionized silica and is most commonly fossilized by calcification (in calcium carbonate) or by pyritization (in iron sulphide) or incorporated (coalified, in varying degree) in peat or lignite. Environmental deductions may be made from the type of siliceous permineralization of the wood (Francis, 1983, 1984, (this volume)). The various processes of fossilization may preserve the cellular details of the wood so well that growth-ring studies may be carried out upon it exactly as if it were modern wood (Creber, 1977; Jefferson, 1982; Francis, 1983). Nevertheless, the approach to the study of growth rings in woods up to many millions of years old differs in a number of important respects from that of historical material at most a few thousand years old.

Most records of fossil wood are of fragments which may only show a limited number of growth rings. Even when 'fossil forests' are found in the site of growth (Jefferson, 1982; Francis, 1983) there is little to be gained by attempting to cross-date the stumps. The principal interest in such material is to detect a general indication of the climatic environment under which the trees grew, rather than to identify fluctuations even of the order of centuries or millennia. This is because the identity of any wood over a few million years old is at best only to be expressed at the level of a living or fossil genus, so that no knowledge of its precise response to the climatic environment is available.

Furthermore, single small pieces of fossil wood make up the bulk of our record and their position within the tree may be only imperfectly established. The site of growth and its topographic relations may be equally unknown.

Despite these limitations, ancient fossil woods represent an important source of palaeoclimatic information which as yet has been little exploited. Even though this may appear crude or blurred to the dendrochronologist or dendroclimatologist of the historical period, it does constitute a source of information on the level of biological productivity and especially on the character of the seasonality (or lack of it) in the palaeoclimatic regime.

OCCURRENCES OF FOSSIL WOOD

1: The Devonian

Fossil wood first appears in the geological record about 370my BP. The earliest trees, (the wood is assigned to the genus Callixylon, the foliage to Archaeopteris) (Figure 1), could be very substantial, with stump diameters in excess of 1m. It is interesting to note that they had evolved only some 30my after the earliest vascular plants (with primary xylem) recorded in the fossil record (Chaloner and Sheerin, 1979). Details of the principal localities where these fossil trees occur are given in Table 1 and Figure 2. In some specimens the wood has no growth rings while in others the ring boundaries are very faint and in irregular sequences. Many ring boundaries appear not to encircle the entire trunk (Lemoigne et al., 1983).

2: The Carboniferous

During the later part of this period from about 330my BP to 250my BP a large area of lowland swamp forest extended from present-day Kansas eastwards across the Mid-West of the United States to the east coast and from Western Europe to the Donetsk Basin in the USSR (Figure 3) (the so-called Euramerian province). The peat-accumulating vegetation included many arborescent species, such as Lycopods (club-mosses), articulates (horse-tails) and seedferns which produced extensive secondary wood and attained the stature of trees. Thus a very large quantity of wood was produced. Studies of the fossilised remnants again show that much is totally devoid of growth rings.

However, in wood specimens drawn from the entire Carboniferous period some growth rings are recorded (Figure 3, Table II) which occur rather haphazardly in both space and time. In general, the climatic conditions apparently permitted much continuous wood growth, but the occurrence of, say, fluctuations in the availability of water caused interruptions in the formation of wood. Some fossil wood recorded from the Carboniferous shows strongly developed growth rings.

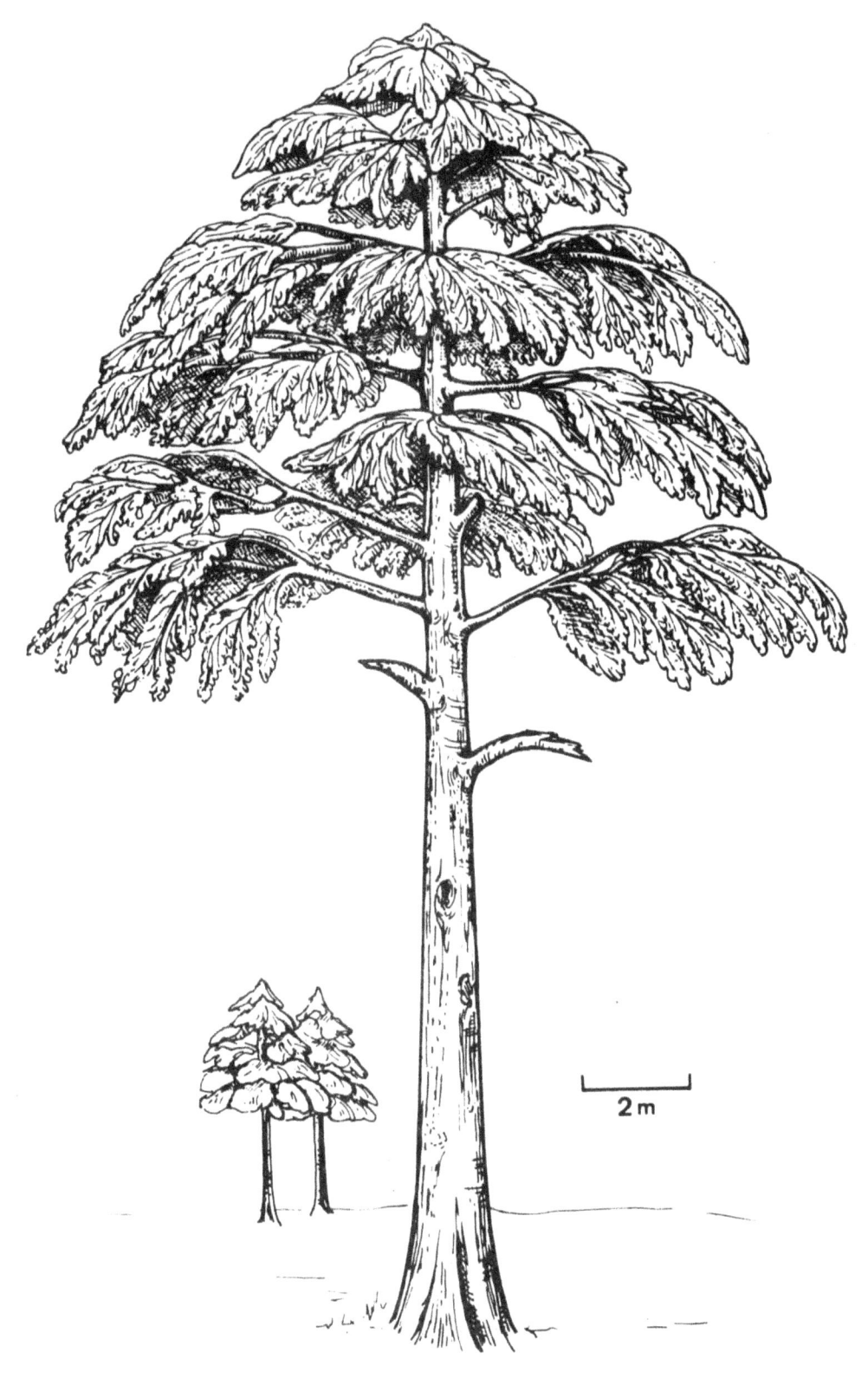

Figure 1. A generalised reconstruction of a member of the genus <u>Archaeopteris</u>, the earliest known genus of trees, from the Upper Devonian (ca. 370my BP). Note the considerable dimensions of the trunk. (After Beck, 1962).

Table I. Upper Devonian fossil woods plotted on **Figure** 2

GENUS	LOCALITY	MAP SITE	PALAEO-LATITUDE	AUTHORS
Callixylon	Indiana USA	1	19°S	Elkins & Wieland,1914 Arnold,1931
Callixylon	New York State USA	2	17°S	Hylander,1922 Arnold,1930 Beck,1953
Sphenoxylon	"	2	17°S	Matten & Banks,1967
Callixylon	Ellesmere Island	3	16°N	Andrews et al,1965
Indet. gymnosperm wood	Spitzbergen W. Germany	4 5	20°N 7°S	Høeg,1942 Kräusel & Weyland,1937
Callixylon	Timan USSR	6	18°N	Lemoigne et al,1983
Callixylon	Donetsk Basin USSR	7	3°N	Zalessky,1909,1911
Callixylon	Kazakhstan USSR	8	3°N	Lepekhina,1963
Callixylon	Tannu Tuva USSR	9	39°N	Lemoigne et al,1983
Palaeo-spiroxylon	Karaganda USSR	10	26°N	Iurina & Lamoigne,1972

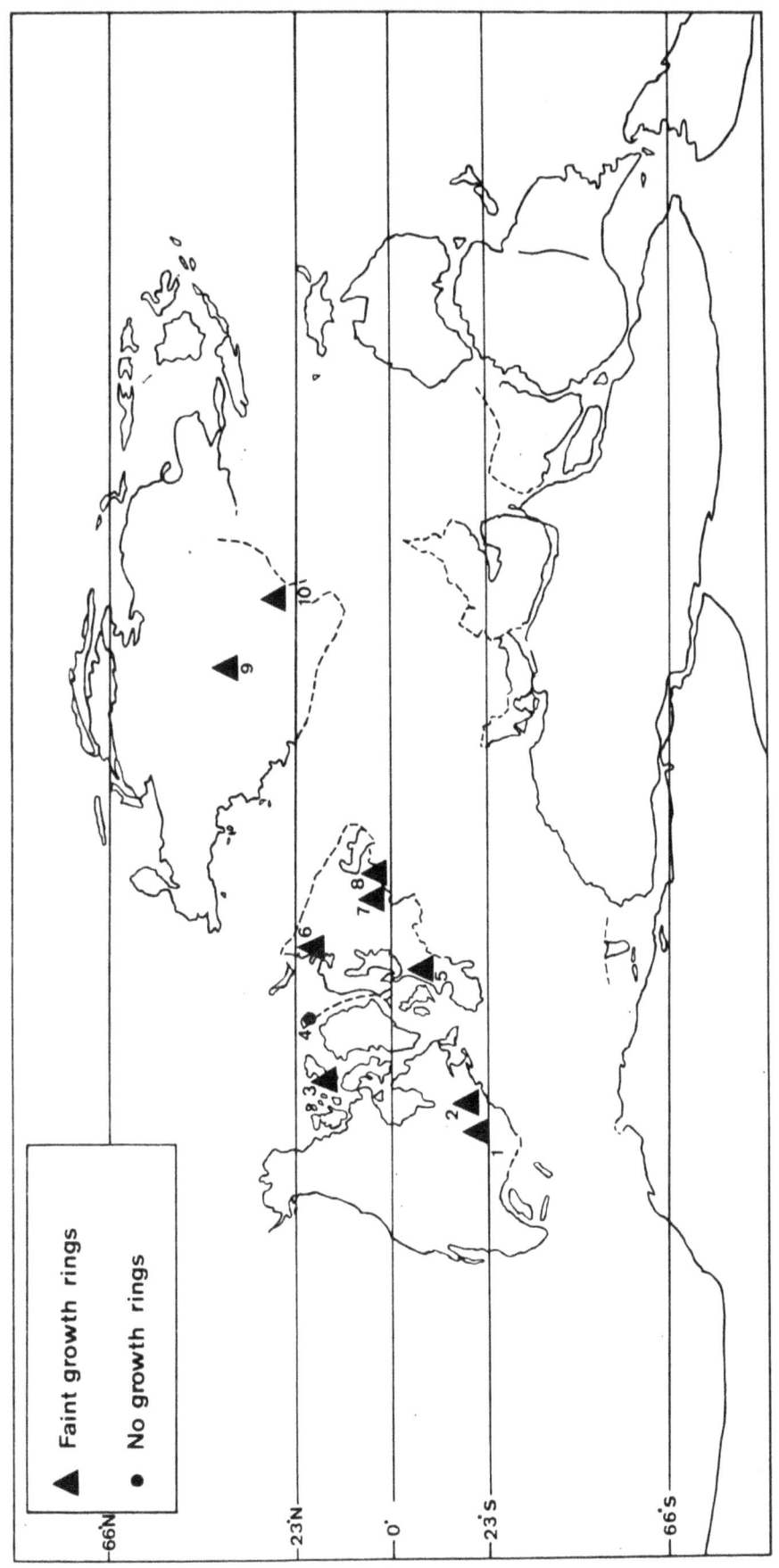

Figure 2. A palaeoreconstruction map of the Upper Devonian Period (ca. 370 my BP; Smith et al., 1981), details of the sites at which fossil wood has been recorded are given in Table I. Note that most of the sites are in the low palaeolatitudes.

3: The Permian

In the early Permian (Figure 4) there was widespread glaciation in the southern hemisphere but the climate there seems to have ameliorated by Upper Permian times so as to permit the growth of trees at very high southern latitudes (No. 8, Figure 4). Moreover the rate of productivity involved in the tree growth at these sites seems to have been very considerable. Doumani and Long (1962) report trunks 8m long and 60cm in diameter and in situ stumps are also recorded (Jefferson, 1983). Maheshwari (1972) reviews the published accounts of the wood and shows that they exhibit marked growth rings. Also, he emphasises that the rings are of substantial thickness, up to 1cm or more. The problems of tree physiology raised by the growth of forests so near to one of the earth's poles are discussed later.

Growth rings are generally not developed in the woods of the Euramerian Province (Gothan, 1911) although Frentzen (1931), Lemoigne and Tyroff (1967) and Schweitzer (1968) have reported some rare exceptions (Table III). However, Permian woods from the Kuznetsk Basin show growth rings (Lepekhina, 1972), some of them strongly developed. In South America, South Africa and India (Nos. 2-7, Figure 4) the Permian trees produced weak growth rings. They are assigned mainly to the fossil wood genera Dadoxylon and Araucarioxylon because of their wood's resemblance to that of trees in the modern genus Araucaria which under a variety of environmental conditions produce very weakly developed growth rings with little latewood. It is possible therefore that the particular ring characteristics of these Permian woods may be genetically determined (Creber and Chaloner, 1984).

4: The Mesozoic and Early Tertiary

From palaeobotanical evidence generally, the Mesozoic was a period of great uniformity of floras on a global scale, lacking the strongly latitudinal zonation of the present-day (Batten, 1984). This situation is further emphasized by the distribution and characteristics of the fossil wood recorded from this period and on into the Early Tertiary (Figure 5 and Table IV). A prominent feature of the distribution is the broad zone from about palaeolatitude 32°N to 32°S in which the wood specimens are either without rings or have only poorly defined ones. An exception is in the region that is now part of China and Japan (Sites 15, 16 and 21, Figure 5) where the woods show definite growth rings. In contrast, locality No. 20 at the same palaeolatitude, has fossil wood specimens which are devoid of rings.

The most notable feature of the distribution of the fossil wood of the Mesozoic is its occurrence in very high northern and southern palaeolatitudes. There is no tree growth in the inhospitable climate at the comparable latitudes of the present day. The very wide ring widths exhibited by the fossil wood specimens further show that the productivity of these high latitude forests was substantial. For comparison some modern records of the growth of trees and other woody plants are plotted on Figure 6 and listed in Table V. The ring widths of the modern plants are very small compared with those in the Mesozoic and Early Tertiary at similar high latitudes. The reasons for this are discussed below.

DISCUSSION

From the foregoing, the fossil record shows a variety of tree species

Table II. Carboniferous fossil woods plotted on **Figure 3**

(L.M.=Lower Mississippian, M.P.= Middle Pennsylvanian,
U.P.= Upper Pennsylvanian.)

GENUS	LOCALITY	MAP SITE	PALAEO-LATITUDE	STAGE	AUTHORS
Dadoxylon	Oklahoma	1	7°S	U.P.	Wilson, 1963
Callixylon	Oklahoma	1	18°S	L.M.	Arnold, 1934
Dadoxylon (Cordaites)	Oklahoma	1	7°S	U.P.	Jensen, 1982
Cordaites	Oklahoma	1	7°S	U.P.	Goldring, 1921
Cordaites	Oklahoma	1	7°S	U.P.	Tynan, 1959
Cordaites	Kansas	2	8°S	M.P.	Arnold, 1947
Callixylon	Kentucky	3	17°S	L.M.	Hoskins and Cross, 1951
Dadoxylon	Ohio	4	11°S	M.P.	Arnold, 1947
Eristophyton	S. Scotland	5	2°S	L.M.	Absalom, 1931
Eristophyton	S. Scotland	5	2°S	L.M.	Lacey, 1953
Pitus	S. Scotland	5	2°S	L.M.	Long, 1979
Taxopitys	E. Siberia	6	49°N	U.P.	Shilkina, 1960

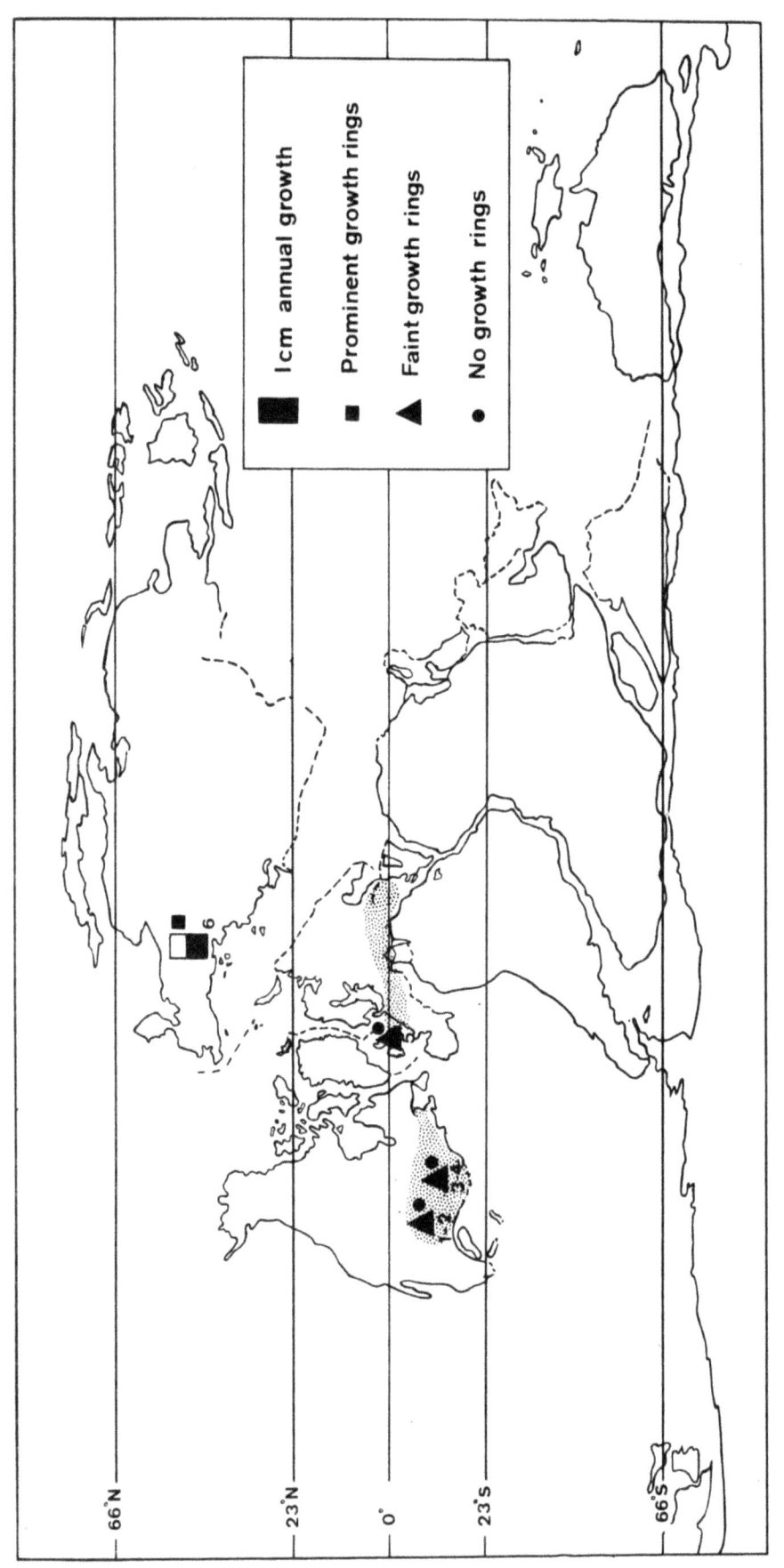

Figure 3. A palaeo-reconstruction map of the mid-Carboniferous Period (ca. 320my BP; Smith et al., 1981), the lowland swamp forest region is shown stippled. Details of the fossil wood sites are given in Table II. As in the Upper Devonian most of the sites are in the low palaeolatitudes.

with differing ring characteristics and variable growth rates. These features require some explanation.

With respect to the Devonian trees, Antevs (1925, 1953) suggested that they had not evolved a mechanism for producing growth rings. Although one might perhaps argue that in the Upper Devonian there was not the complex of plant growth regulators that exist at the present day, it is difficult to conceive of a plant structure as large as a tree being able to co-ordinate its development without them. Furthermore, Antevs' suggestion does not apply to ringless trees of later geological Periods.

Although genetic characteristics may be the dominant factor accounting for the ring patterns of some trees such as Araucarioxylon mentioned from the Permian, this cannot be a general explanation, particularly where modern similar trees are lacking. The most useful mechanism for explaining these patterns is plate tectonics, and the movements of the continents in relation to the main climatic belts. It is important to place fossil woods in their palaeolatitudes, rather than their present day latitudes when found. The past position of continents have now been reasonably well established, principally from the evidence of palaeomagnetism (Tarling, 1983). It can be shown that the ringless, or weakly ringed trees of Devonian times grew close to the Devonian palaeo-equator, or within the sub-tropical belt of those times. Antevs' explanation thus becomes unnecessary. Similary the Euramerian province of Carboniferous and Permian times was in palaeo-equatorial latitudes, and show wood with no, or weakly developed rings, for example the Tylodendron stem described by Holden (1913) from Prince Edward Island, virtually on the Permo-Carboniferous palaeo-equator, which was completely lacking in rings. In contrast, the only fossil wood recorded from a Carboniferous high palaeolatitude (No. 6, Figure 3) shows strongly developed growth rings. The ringed woods from the Kuznetsk Basin grew at a palaeolatitude of 29°S, whilst the very high latitude Permian wood had clear rings (Maheshwari, 1972).

On this basis, the widespread occurrence of wood with little or no rings during the Mesozoic and Early Tertiary between palaeolatitudes 32°N to 32°S would suggest a much broader belt of almost or entirely seasonless climate than is found at the present day, and this has been supported independently on other grounds by Hallam (1984).

Finally, the anomalous occurrences of clearly defined rings in China and Japan at this time, and the unusual absence of rings at locality 20, also fit a climatic explanation. Computer modelling by Barron and Washington (1982) predicted large climatic variations in exactly this region, and this matched the ring formation in the trees. At the same time, Kimura (1984) has shown that the Mesozoic flora at locality 20 is quite distinct from that of the rest of Japan and it is thought that the region is a displaced terrane (tectonic block) which has been moved from much farther south. Given such a lower latitude, the trees could have produced wood lacking in growth rings. This is an instance of the use of growth ring studies in testing certain aspects of palaeocontinental dispositions.

Having established the link between ring characteristics and climate, the remaining problem is to account for the high productivity of trees at high palaeolatitudes compared to that of trees at modern high latitudes. The productivity of trees in the polar regions of the present day may be limited either by the prevailing low ambient temperatures, or by the light regime alone. Warren Wilson's extensive study (1964) of the physiology of

Table III. Permian fossil woods some of which are plotted on Figure 4

SPECIES	LOCALITY	MAP SITE	PALAEO-LATITUDE	AUTHORS
Araucarioxylon spp.	Kuznetsk Basin USSR	–	29°N	Lepekhina, 1972
A. Humiliradiale	"	"	"	"
A. originale	"	"	"	"
Eristophyton zalesskyi	"	"	"	"
Not assigned	Upper Rhine District, W.Germany	–	Approx 13°N	Schweitzer, 1968
Lebachia piniformis	Nahe Basin W.Germany	–	Approx 11°N	Lemoigne & Tyroff, 1967
Tylodendron sp.	Prince Edward Is. E.Canada	1	3°S (LP)	Holden, 1913
Podocarpoxylon sp.	Bagé S.Brazil	2	41°S	Leistikow & Creber
Various spp.	Reuning's Fossil Forest Franzfontein S.W. Africa	3	42°S	Kräusel & Range, 1928
Dadoxylon bakeri	Falklands Islands	4	55°S	Seward & Walton 1923
Palaeospiroxylon heterocellularis	Bengal, India	5	39°S (UP)	Prasad & Chandra 1980
Dadoxylon indicum	Bengal, India	6&7	62°S (LP)	Holden, 1917
D. bengalense	"	"	"	"
Various spp.	Trans-antarctic mountains	8	70°S	Maheshwari, 1972

LP = Lower Permian
UP = Upper Permian

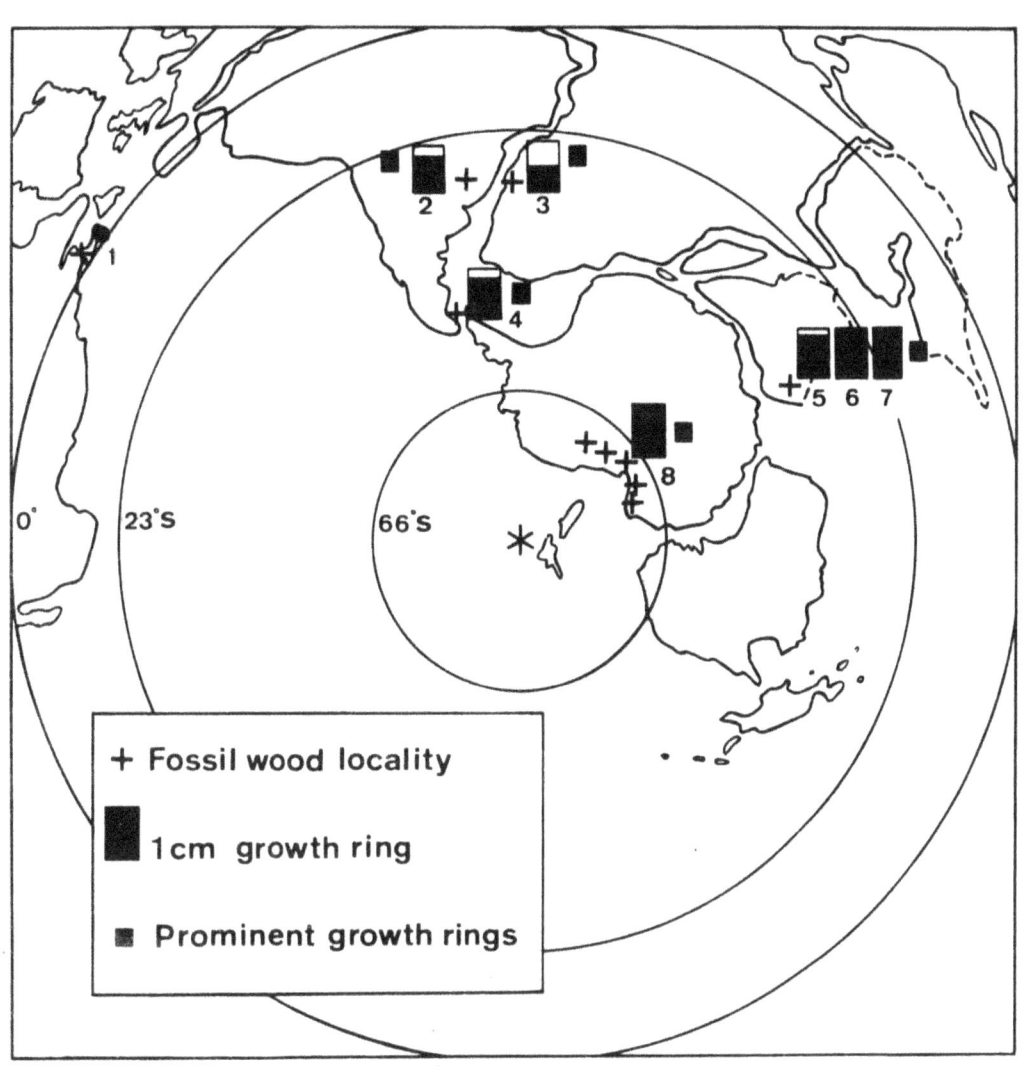

Figure 4. A palaeo-reconstruction map of a south polar view of the world in the Upper Permian Period (ca. 240my BP; Smith et al., 1981). Details of the fossil wood sites are given in Table III. Note that wood specimens with wide rings are located in very high palaeolatitudes.

arctic plants, showed that the mean annual radial growth of Salix arctica was only 0.07mm on Cornwallis Island (75°N) whereas in southern Alaska (60°N), Cooper (1931) found that for Alnus tenuifolia it was 2.73mm. Elkington and Jones (1974) provided further evidence of this kind by showing that Betula pubescens s. lat. in southwest Greenland (61°N) had only 25% of the annual productivity of Betula in England (Ovington and Madgwick, 1959). The other examples of ring widths in the Arctic at the present day plotted on Figure 6 all tend to support the general pattern of low productivity (as expressed in wood growth increments) in high latitudes.

There is considerable evidence to support the view that low ambient temperature is the dominant control. For example, it is significant that the northern limit for coniferous tree growth is almost coincident with the 10°C July isotherm. Furthermore, Oswald (1969) has shown that temperature controls the annual timber increment of Picea abies in the Central Massif, France. At timberline (1650m), the increment is one-ninth of that at 1200m elevation, and this reduction is attributed to the low ambient temperature at the beginning and end of the growing season. By analogy, the low temperatures in high latitudes at the present day have reduced the growing season for trees to zero.

A third indication comes from the work of Warren Wilson (1966), who drew attention to the significant fact that although the mean summer temperature on Cornwallis Island (75°N) is only 4°C the June solar input is barely less than that for Washington D.C. in the same month (Table VI). His work in arctic plant physiology suggested that the low ambient temperature limited those metabolic processes involved in the transport and use of photosynthate very much more than the process of photosynthesis itself. Thus assimilates tended to accumulate and their concentration in leaves rose to a level which depressed assimilation to a rate which roughly balances their rate of use. Russell (1940) and Warren Wilson (1966) have both shown that the sugar concentration in the leaves of such arctic plants as Ranunculus glacialis, Oxyria digyna and Salix herbacea is two to four times the concentration in comparable leaves of plants in warmer climates. Further data on mean June temperatures and the solar radiation rate in high northern latitudes are given in Table VI.

Fourthly, more recently, Tieszen (1978), when describing experiments investigating the effects of temperature variations on the rate of photosynthesis in Dupontia fisheri provides additional support for the view that growing conditions for plants in the present day Arctic are limited by the low ambient temperatures and not by an inadequate light supply. He says "the seasonal course of CO_2 uptake was surprisingly insensitive to the temperature changes providing further support for the generalization that photosynthesis per se is not strongly reduced by the low arctic temperatures" (p. 626). He continues by saying "leaf growth or the allocation for leaf production is temperature-sensitive and may be much more significant than the temperature sensitivity of photosynthesis in limiting primary production" (p. 626).

A final piece of evidence is provided by Axelrod (1984), who recently presented data from a wide range of plants. He describes the way in which more than a hundred species of tropical plants may be induced to grow, flower and fruit under greenhouse conditions near Stockholm (59°N), and, remarkably, no arrangements for artifical illumination have to be made.

Whilst Axelrod's work shows that lower latitude plants may grow in

Table IV. Jurassic, Cretaceous and Lower Tertiary fossil woods plotted on **Figure 5**

AGES:- J Jurassic, L.C. Lower Cretaceous, L.J. Lower Jurassic, L.T. Lower Tertiary, M.C. Middle Cretaceous, M.J. Middle Jurassic, U.C. Upper Cretaceous, U.J Upper Jurassic

RING DETAILS:- A Absent, F Faint, L Largest

LOCALITY	AGE	MAP SITE	PALAEO-LATITUDE	RING DETAILS	AUTHORS
Northern Alaska	L.C.	1	80°N	L 4.0mm	Arnold, 1952
Amund Ringnes Is.	L.C.	2	75°N	L 6.5mm	Bannan & Fry, 1958
Ellesmere Is.	L.T.	3	74°N	L 3.0mm	Christie, 1964 Bradley, 1982 (pers. comm)
Spitzbergen King Charles Land	L.T.	4	61°N	L 4.4mm	Gothan, 1907
Hare Is. West	L.T.	5	62°N	L 5.4mm	Creber, not published
Chilko Lake B.C.	L.C.	6	59°N	L 5.6mm	Fry, 1958
Rosedale Alberta	U.C.	7	51°N	L 1.8mm	Ramunajam & Stewart, 1969
Dakota USA	L.C.	8	48°N	L 1.25mm	Read, 1932
East Sutherland	U.J.	9	43°N	L 9.0mm	Creber, 1972
Sonora Mexico	L.C.	10	35°N	L 7.5mm	Cevallos-Ferriz, 1984 Cevallos-Ferriz & Gonzalez-Leon, 1983
Coahuila Mexico	U.C.	11	30°N	L 10.0mm	Cevallos-Ferriz, 1984 Cevallos-Ferriz & Weber (unpubl.)
N. Carolina USA	U.C.	12	28°N	F	Boeshore & Gray, 1936
South Portugal	U.J.	13	31°N	A	Boureau, 1949
Dorset	U.J.	14	36°N	L 3.7mm	Francis, 1983
Hopeh China	U.J.	15	37°N	L 5.0mm	Chang, 1929

Table IV. Continued

Location	Type	No.	Lat.	Size	Reference
Liaotung, Manchuria	L.C.	16	35°N	L 2.5mm	Shimakura, 1937
Pyongyang	M.J.	17	43°N	L 13.0mm	Shimakura, 1937
Koryak, USSR	U.J.	18	72°N	L 3.0mm	Shilkina, 1963
S. Sakhalin Japan	U.C.	19	56°N	L 3.7mm	Shimakura, 1937
Chiba, Japan	L.C.	20	36°N	A	Nishida, 1973
Koti-ken Japan	U.J.	21	36°N	L 3.0mm	Shimakura, 1936
Nong-son Vietnam	L.J.	22	28°N	F	Boureau, 1950
Changwat Kalasin Thailand	L.J.	22	28°N	F	Asama, 1982
Soegi, Indon.	L.J.	23	21°N	A	Roggeveen, 1932
Afghanistan	J.	24	22°N	F	Sitholey, 1940
Uzbekistan USSR	U.C.	25	32°N	F	Khudayberdyev, 1962
Morocco	U.J.	26	24°N	F or A	Boureau, 1951
Tripolitania	M.C.	27	13°N	F or A	Negri, 1914
Niger	L.C.	28	2°N	F	Williams, 1930
Cairo	L.T.	29	5°N	F or A	Unger, 1859
Tchad	L.C.	30	7°S	F	Boureau, 1952
Tanzania	U.C.	31	28°S	F or A	Potonié, 1902
Madagascar	U.C.	32	31°S	A	Fliche, 1900
Bihar, India	M.J.	33	40°S	L 1.2mm	Kräusel & Jain, 1963
E. Pondo Land South Africa	U.C.	34	48°S	F or A	Mädel, 1960 Schultze-Motel, 1966
Santa Cruz Argentina	U.J.	35	53°S	L 2.5mm	Calder, 1953
Alexander Is. Antarctica	L.C.	36	70°S	L 9.0mm	Jefferson, 1982
S. Island N.Z.	M.J.	37	70°S	L 1.1mm	Pole, 1982
S. Island N.Z.	U.C.	38	67°S	L 1.0mm	Stopes, 1916

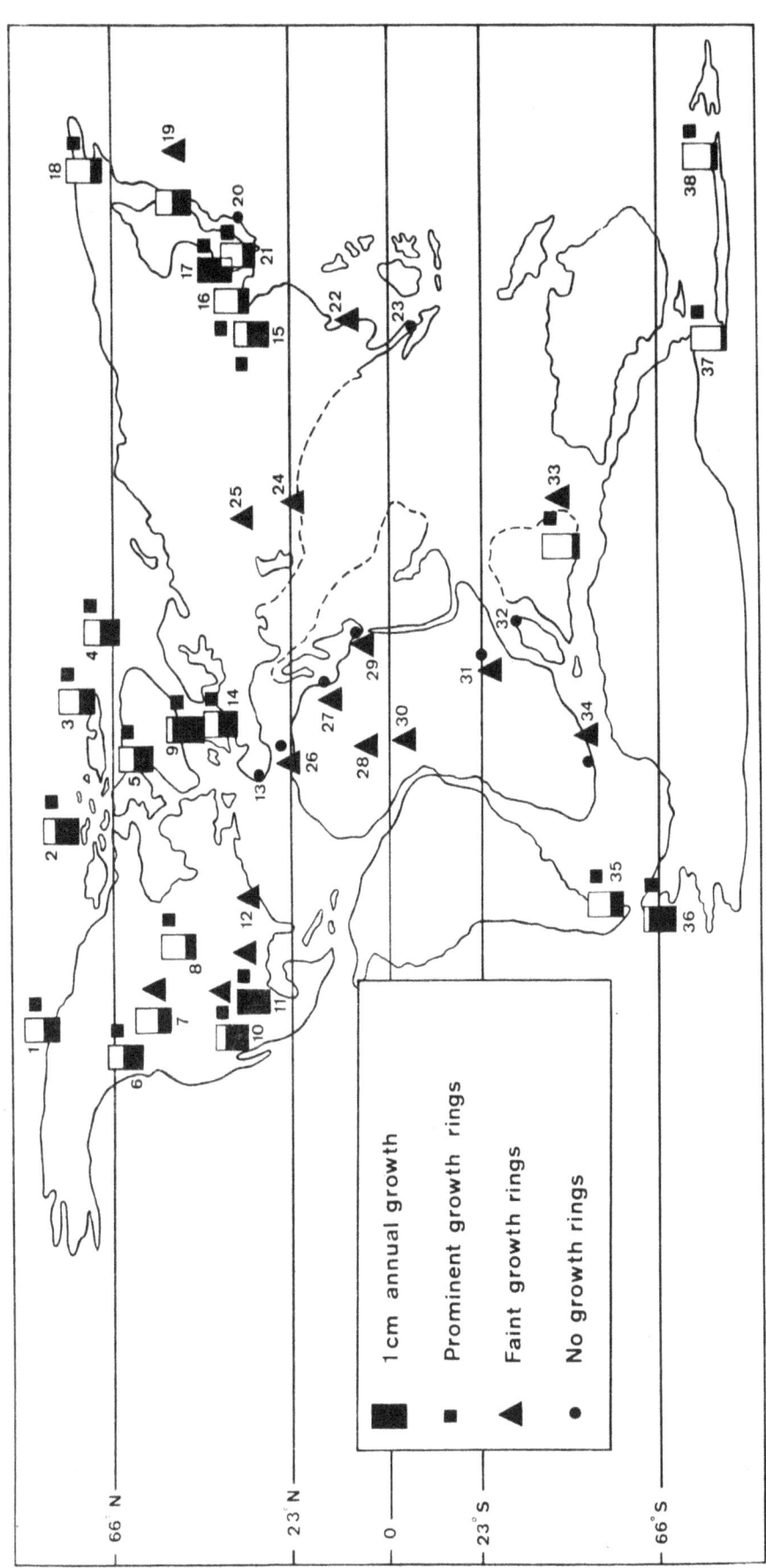

Figure 5. A palaeo-reconstruction map of the Lower Cretaceous Period (ca. 120my BP; Smith et al., 1981). Details of the fossil wood sites are given in Table IV. Note the broad zone in the low palaeolatitudes where growth rings are absent or only weakly defined and also the number of sites in the higher palaeolatitudes.

higher latitudes, if they are provided with the necessary ambient temperatures, tree growth in high latitudes still requires adequate light and the ability to intercept it at low angles without too much mutual shading of adjoining trees. As Frakes (1979) points out, in discussing the high palaeolatitude coals of New Zealand and Alaska, "the main problem that arises from these distributions has to do with the availability of sunlight for plant growth. It has been stated by many workers that the limitation on the existence of plants at high latitudes is the sunlight factor" (p. 176). It is possible to present evidence that the solar input at very high latitudes may very well be adequate for the growth of forests.

The solar input in Antarctica appears to be just as adequate for tree growth as it is in the Arctic. At two locations well inside the Antarctic Circle (LeGrange, 1963: 82°S; Farman and Hamilton, 1978: 75.5°S) the total solar energy input was 3664 MJ $m^{-2}y^{-1}$ and 3452 MJ $m^{-2}y^{-1}$ respectively; both of these figures are larger than the total annual input for East Anglia (3212 MJ $m^{-2}y^{-1}$, Ovington, 1961). It would thus appear that there would be no growth-limiting shortage of light energy in the high southern latitudes. However, the timing of this light energy is clearly different in high latitudes, with very long daylight hours during the growing season, which is very short. As regards the long daylight hours, Vaartaja (1962) has demonstrated the existence of what he terms "photoperiodic ecotypes" in a wide variety of modern tree species which only produce their best growth in their natural habitats in the higher latitudes. Attempts to grow them in shorter photoperiods (i.e. in lower latitudes) result only in stunted growth. Wassink and Wiersma (1955) had previously shown a similar effect with a high latitude provenance (66°) of Pinus sylvestris. In the Mesozoic and Early Tertiary the very high latitude forest species may have had similar photoperiodic ecotypes adapted to temperate, high latitude environments which no longer exist.

Regarding the growing season, although in very high latitudes it is relatively short, the work of Ford et al (1978) suggests that it could have been adequate in the past. They showed that in Sitka spruce the xylem mother cells were produced by the cambial initial cell in a file of cells at the rate of four per day when the light energy input was 22.5MJ per day. At very high latitudes at the present day the light input is of this order for three months at least. Thus 360 cells of about 40 m radial diameter could be formed to produce a total ring width for the season of 14.40mm. This rate of production is clearly adequate to form even the largest ring widths normally met with in forest trees in the most favourable environments.

Conifers growing in high latitudes at the present day show a further adaptation that enables them to produce adequate amounts of wood in the short growing season. Gregory and Wilson (1968) showed that the production of xylem cells by the cambium of Picea glauca, white spruce, growing near College, Alaska (65°N) was at a much higher rate than in trees of the same species growing at Petersham, Massachusetts (42°N). In considering the length of the growing season it is evident that many tree species do not "use" the whole of the optimal periods available to them. Kramer (1943) showed that Pinus resinosa only requires 140 days and Abies balsamea and Pinus strobus 120 days to complete their seasonal growth. In very high latitudes the illuminated part of the year could well have been adequate as a full growing season in the geological past. This fact is further emphasized by Gosz et al. (1978) who show that a complete, temperate deciduous forest ecosystem is supported on only the 2016MJ of light energy

Table V. Recent ring width data from high latitudes at the present day, plotted on **Figure** 6

SPECIES	MAP SITE	RING WIDTH (mm)	LOCALITY	LATITUDE (°N)
Larix dahurica	A	W2.0	River Novaja	72
" "	A	W1.9	River Boganida	71
" "	A	W5.0	Between Amginsk & the River Aldan	60
Pinus sylvestris	B	W2.1	Leningrad	60
" "	B	W9.0	Estonia	59
Pseudotsuga menziesii	C	W5.0	British Columbia	55
Betula nana	D	A0.14	Kaiser Franz Joseph's Fjord, East Greenland	73
" "	D	A1.6	Wurzburg, Germany	48
Vaccinium uliginosum	E	A0.032	Kaiser Franz Joseph's Fjord, East Greenland	73
" "	E	A0.7	Erlanger, Germany	49
Alnus tenuifolia	F	A2.7	Glacier Bay Alaska	59
Salix arctica	G	A0.07	Cornwallis Island	75
" "	H	A0.2	Axel Heiberg Island	79
Betula pubescens	I	A0.5	S.W. Greenland	61

W=widest ring width A=average ring width

The above data was obtained from : Middendorf, 1867 (A & B); British Columbia Forest Service, 1947 (C); Kraus, 1874 (D & E); Cooper, 1931 (F); Warren Wilson, 1964 (G); Beschel & Webb, 1963 (H); Elkington & Jones, 1974 (I).

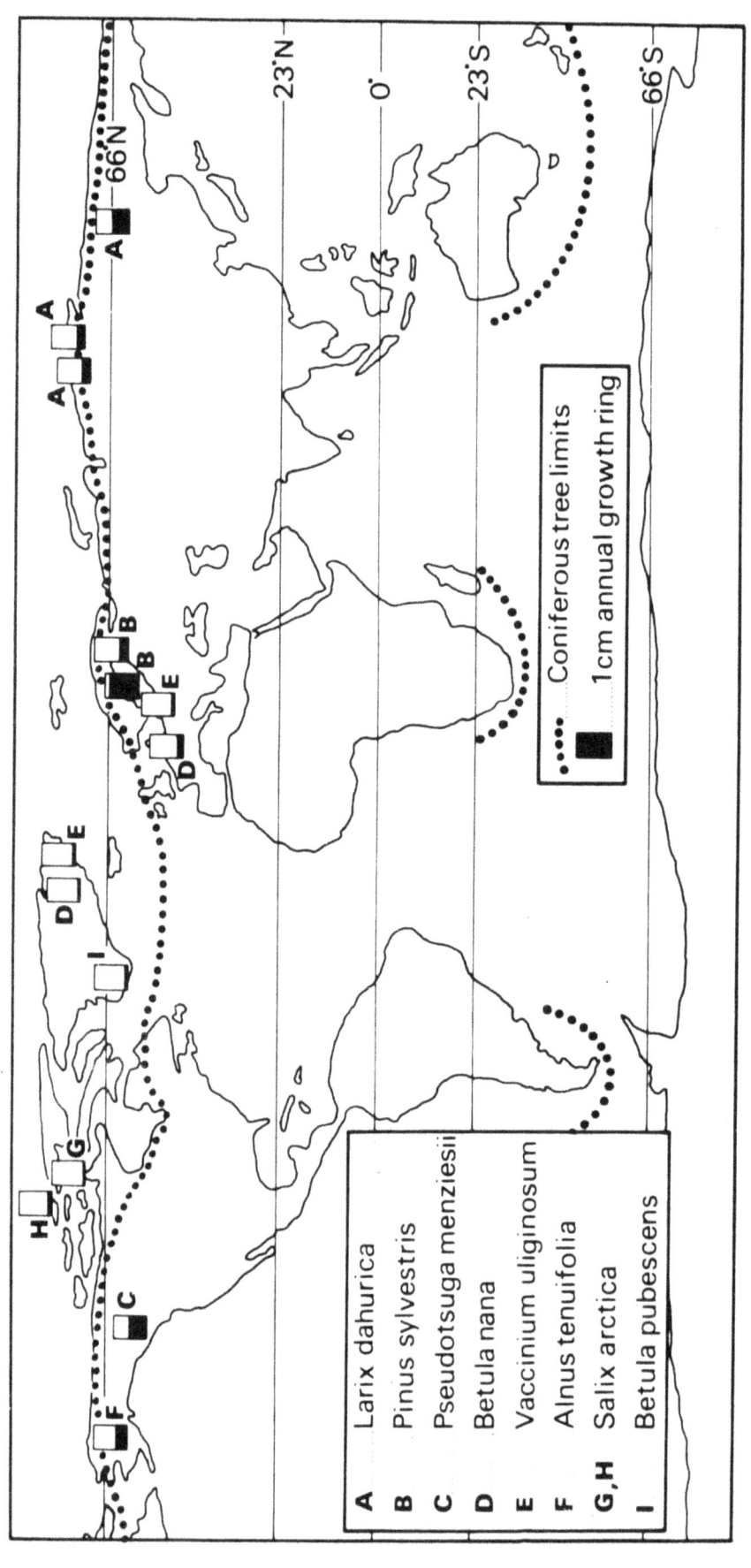

Figure 6. A map showing the distribution of the continents in the Recent world. Note the very small growth increments in the high latitudes in sharp contrast with those of the Mesozoic. Details of the lettered sites are given in Table V.

that falls on the trees in the few months of the year that they are in leaf. This quantity is substantially less than that received within the Arctic Circle during the summer. Unfortunately we have little information about the leaf-shedding strategy of many of the high-latitude fossil trees of the Late Cretaceous-Early Tertiary. Obviously, as in Larch, a deciduous habit may have been part of a high-latitude overwintering strategy. However, there is no evident correlation between deciduous habit and distribution into high latitudes among living conifers.

A final problem relates to the low angle of incidence of the sun's rays in high latitudes, which causes heavy shadow. A corollary of this is that trees require adequate spacing and suitable crown shapes to make maximum use of the light available.

In this context the detailed description of a fossil forest by Jefferson (1982) is of interest. This forest grew during the Lower Cretaceous at a palaeolatitude of about 70°S and was apparently overwhelmed in a single event. In one continuous geological exposure there were 31 coniferous tree stumps in an area of 550m², giving a minimum density of one tree per 17m². In forestry terms these are rather widely spaced at 588 trees ha^{-1}, since comparable living trees are capable of growing at a density of 1000 trees ha^{-1} in forest stands at lower latitudes.

To calculate the area of crown presented to the incoming solar radiation, the formula of Jahnke and Lawrence (1965) may be used for conically shaped trees of various dimensions and for a range of angles of the sun above the horizon. This shows that at low angles of elevation, a cone-shaped tree whose crown height is much greater than its basal radius intercepts most light (Figure 7). It is noteworthy that the conifers, that group of tree-forming species which develop such substantial forests at high latitudes, generally show this growth habit; whilst low latitude trees have umbrella shaped crowns which are appropriate for high angles of incidence.

Conical crowns of height 17m² and basal radius 1.6m could have been borne by Jefferson's trees without the lower branches touching. Such crowns have light-intercepting areas at low solar angles of about 25m². Clearly some allowance for shading must be made but it should be pointed out that in midsummer at latitude 70°S the sun travels through 360° and reaches an elevation of 43.5° above the horizon at midday, thus one tree will not shade another throughout the period of daylight. With a directly illuminated effective surface of 25m² the tree could intercept over 50,000MJ of direct radiation during the illuminated months of the year. With a total conical surface area of 86m² the tree would also be well placed to receive diffuse radiation which, as Monteith (1973) points out, may be up to 25% of the incoming radiation even on a cloudless day.

The annual increment of trunk wood produced by a tree represents only a small part of the total amount of solar energy incident upon its crown (Figure 8). Ovington (1961) estimates that a coniferous forest utilizes about 1% of the incident sunlight and Kozlowski (1962) considers that the resulting photosynthetic product is divided between trunk wood and all other uses in the ratio 40% ; 60%. Jefferson (1982) calculated that the average ring width in his antarctic fossil trees was 2.5mm, which for one tree, would represent about 7.8kg of wood with an energy content of about 164MJ. With a possible light energy flux of 50,000MJ during a year, the 164MJ required for wood production in these antarctic trees would have only represented about 0.3% of the total energy flux rather than 0.4% as recorded

Table VI. Solar radiation in high latitudes; daily rate in June.

LOCATION	LATITUDE (°N)	MEAN JUNE TEMPERATURE (°C)	SOLAR RADIATION RATE (MJ m^{-2}day^{-1})
Abisko, Sweden	68	6.3	19.7
Umiat, Alaska	69	8.5	17.2
Resolute, Cornwallis Island	75	4.0	27.8
Greenharbour, Spitzbergen	78	2.4	22.9
Sveanor Spitzbergen	80	2.1	24.4
Lower latitude site for comparison, Washington D.C.	39	23.7	31.5

It should be noted that the receipt of solar radiation per day at all latitudes above the Arctic Circle is enhanced by the fact that there is sunlight throughout 24 hours.

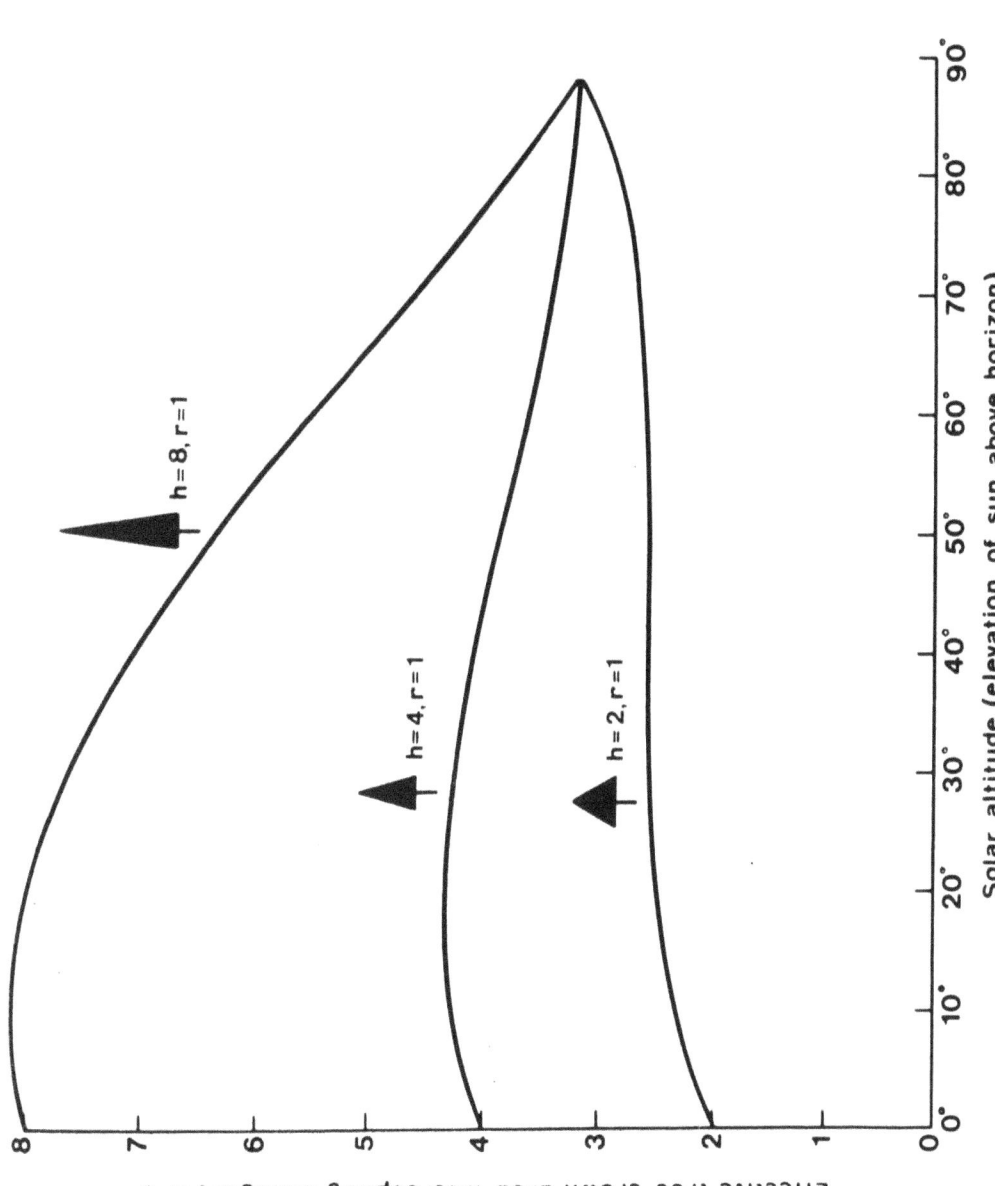

Figure 7. A family of curves showing the effective surface areas for the absorption of solar radiation on conical tree crowns of three different ratios of height to basal radius. The curves show that for low angles of incident sunlight, a tall tree crown intercepts disproportionately more light than a crown of the same basal radius but lesser height. With sunlight incident vertically upon the earth (at 90°) the effective area of light absorption is 3.14m^2 (πr^2) for all three crowns illustrated regardless of height. h: height; r: radius (after Jahnke and Lawrence, 1965).

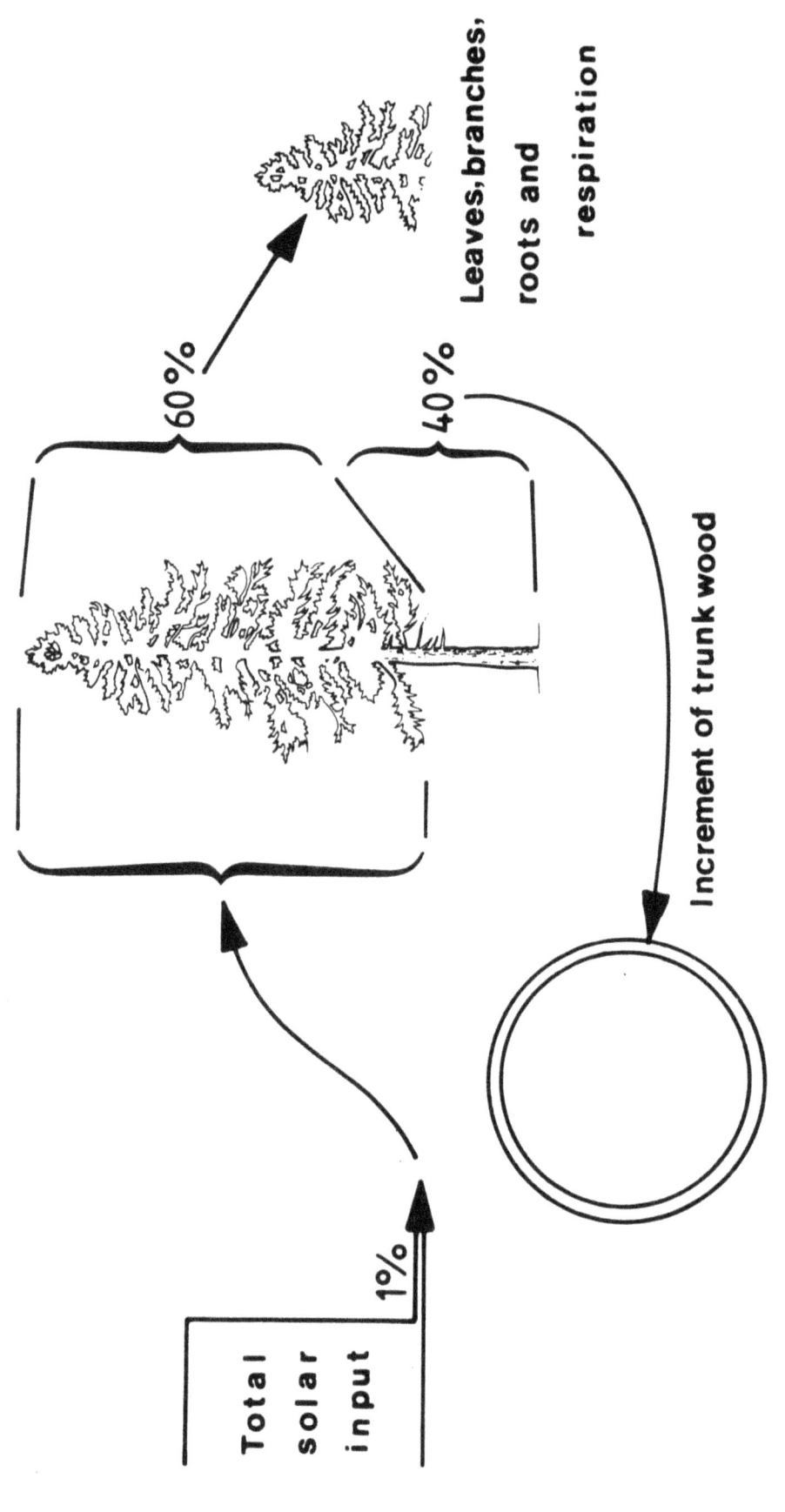

Figure 8. The ultimate fate of solar radiation absorbed by the crown of a tree. Only about 1% of the total solar input is utilised and this is apportioned approximately in the ratio 40% : 60% between trunk wood and all other uses.

by Ovington and Kozlowski for temperate latitudes. Losses due to cloud cover and mutual shading of trees can thus be accommodated within these figures. With a production of 7.8kg of wood on trees spaced at one per 17m² this represents an annual timber increment of 0.46kg m^{-2}; thus a productivity of nearly half a kilogram of wood per square metre would have been quite possible in the antarctic light regime.

In summary therefore, at present there is a perfectly adequate supply of light energy at high latitudes to support tree growth, and this can be assumed to be true of past geological periods as well. The main limiting factor on tree growth at the present day appears to be the low ambient temperature.

CONCLUSION

We have shown that the fossilised tree remains from earlier geological Periods are found in regions which were largely in low palaeolatitudes at the time of growth. These remains are either without growth rings or possess rings that are very weakly developed, characteristics which are exactly what would be expected in trees growing in low latitudes. In later geological Periods (the late Palaeozoic and Mesozoic) the main interest lies in the vigorously growing trees in high polar latitudes. Indeed the present latitudinal restriction of tree growth may represent an atypical situation, like the occurrence of polar ice caps which has taken place only infrequently throughout earth history. We have presented elsewhere (Creber and Chaloner, 1985) an hypothesis to account for the occurrence of the global climatic conditions which would permit trees to flourish much closer to the poles than is the case now. In constructing this hypothesis we see no need to invoke anything other than an increase in ambient temperature in the polar regions, since no restriction on growth need arise from the availability of light energy.

Acknowledgements

Thanks are due to the Leverhulme Trust for their generous support of this research. We are much indebted to the following for access to fossil wood specimens: Trustees of the British Museum (Natural History); the Director of the Senckenberg Institute, Frankfurt-am-Main and the Director of the Komarov Institute, Leningrad. Data were very kindly made available to us through the kindness of the late Dr. T.H. Jefferson (Alexander Island, Antarctica) and Dr. J.E. Francis (Basal Purbeck, Britain). In addition, Dr. R.S. Bradley presented to us some fossil wood from Ellesmere Island in the Canadian Arctic. The following gave us permission to reproduce figures: Figure 1 (American Journal of Botany); Figure 6 (J. Wiley and Sons); Figure 7. (Botanical Review). We also thank the many colleagues who gave us helpful advice.

REFERENCES

Absalom, R.G. 1931. Calamopitys (Eristophyton) beinertiana containing annual rings. Northwest Naturalist 6, pp. 70-74.

Andrews, H.N., Phillips, T.L. and Radforth, N.W. 1965. Paleobotanical studies in Arctic Canada I. Archaeopteris from Ellesmere Island. Canadian Journal of Botany 43, pp. 545-556.

Antevs, E. 1925. The climatologic significance of annual rings in fossil woods. American Journal of Science 5th Series 9, pp. 296-302.

Antevs, E. 1953. Tree rings and seasons in past geological eras. Tree-Ring Bulletin 20, pp. 17-19.

Arnold, C.A. 1930. The genus Callixylon from the Upper Devonian of central and western New York. Papers of the Michigan Academy of Science 11, pp. 1-50.

Arnold, C.A. 1931. On Callixylon newberryi (Dawkins) Elkins et Wieland. Michigan University Museum of Paleontology Contributions 3, pp. 207-232.

Arnold, C.A. 1934. Callixylon whiteanum sp. nov., from the Woodford chert of Oklahoma. Botanical Gazette 96, pp. 180-185.

Arnold, C.A. 1947. An introduction to paleobotany. McGraw-Hill Book Co. New York.

Arnold, C.A. 1952. Silicified plant remains from the Mesozoic and Tertiary of Western North America II. Some fossil woods from Alaska. Papers of the Michigan Academy of Science 38, pp. 9-20.

Asama, K. 1982. Contributions to the geology and palaeontology of south-east Asia CCXXII: Araucarioxylon from Khorat, Thailand. Geology and Palaeontology of Southeast Asia 23, pp. 57-64.

Axelrod, D.I. 1984. An interpretation of Cretaceous and Tertiary biota in polar regions. Palaeogeography, Palaeoclimatology, Palaeoecology 45, pp. 105-147.

Bannan, M.W. and Fry, W.L. 1957. Three Cretaceous woods from the Canadian Arctic. Canadian Journal of Botany 35, pp. 327-337.

Barron, E.J. and Washington, W.M. 1982. Cretaceous climate: A comparison of atmospheric simulations with the geologic record. Palaeogeography, Palaeoclimatology, Palaeoecology 40, pp. 103-133.

Batten, D.J. 1984. Palynology, climate and the development of Late Cretaceous floral provinces in the Northern Hemisphere; a review. In: Fossils and Climate (ed. Brenchley, P.J.) Chichester: Wiley and Sons, pp. 127-164.

Beck, C.B. 1953. A new root species of Callixylon. American Journal of Botany 40, 226-233.

Beck, C.B. 1962. Reconstructions of Archaeopteris, and further considerations of its phylogenetic position. American Journal of Botany 49, pp. 373-382.

Beschel, R.E. and Webb, D. 1963. Growth ring studies on arctic willows. In: Axel Heiberg Island, Preliminary Report 1961-1962. (ed. Muller, F.) Montreal: McGill University, pp. 189-198.

Boeshore, I. and Gray, W.D. 1936. An Upper Cretaceous wood; Torreya antiqua. American Journal of Botany 23, pp. 524-528.

Boureau, E. 1949. Dadoxylon (Araucarioxylon) teixeirae n. sp. Bois fossile due Jurassique supérieur portugais. Comunicacoes dos servicos geologicos de Portugal 29, pp. 187-194.

Boureau, E. 1950. Contribution à l'étude paléoxylogique de l'Indochine. Bulletin du Service Géologique de l'Indo-China 29, pp. 1-13.

Boureau, E. 1951. Etude paléoxylogique de l'Afrique du Nord (1): Présence de Dadoxylon (Araucarioxylon) teixeirae Boureau dans le Haut Atlas de Maroc. Notes et Mémoires du Service des Mines et de la Carte géologique du Maroc 4, pp. 123-133.

Boureau, E. 1952. Etude des fossiles du Territoire du Tchad: 1. Protopodocarpoxylon rochii n. sp. Bois fossile mesozoique. Bulletin du Museum d'histoire naturelle de Paris 2° Série 24, pp. 223-232.

British Columbia Forest Service. 1974. Yield Data. British Columbia Forest Service.

Calder, M.G. 1953. A coniferous petrified forest in Patagonia. Bulletin of the British Museum (Natural History) Geology 2, pp. 99-138.

Cavallos-Ferriz, S. and Gonzalez-Leon, C. 1983. Protopodocarpoxylon Eckhold en el Cretacico inferior de Lampazos, Sonora, Mexico. (in preparation).

Cavallos-Ferriz, S. 1984. Description of three Cretaceous gymnosperm woods from Coahuila and Sonora, Mexico. Abstracts of the 2nd International Organisation of Palaeobotany Conference 6.

Chaloner, W.G. and Sheerin, A. 1979. Devonian macrofloras. In: The Devonian System. (eds. House, M.R., Scrutton, C.T. and Bassett, M.G.). Special Papers in Palaeontology 23, pp. 145-161.

Chang, C.Y. 1929. A new Xenoxylon from North China. Bulletin of the Geological Society of China 8, pp. 243-251.

Christie, R.L. 1964. Geological reconnaissance of northeastern Ellesmere Island, District of Franklin. Bulletin of the Geological Survey of Canada Memoir 331, pp. 1-71.

Cooper, W.S. 1931. A third expedition to Glacier Bay, Alaska. Ecology 12, pp. 61-95.

Creber, G.T. and Chaloner, W.G. 1984. Influence of environmental factors on the wood structure of living and fossil trees. The Botanical Review 50, pp. 357-448.

Creber, G.T. and Chaloner, W.G. 1985. Tree growth in the Mesozoic and the Early Tertiary and the reconstruction of palaeoclimates. Palaeogeography, Palaeoclimatology, Palaeoecology 52, 35-60.

Doumani, G.A. and Long, W.E. 1962. The ancient life of the Antarctic. Scientific American 207(3), pp. 168-184.

Elkins, M.G. and Wieland, G.R. 1914. Cordaitean wood from the Indiana black shale. American Journal of Science 38, pp. 65-78.

Elkington, T.T. and Jones, B.M.G. 1974. Biomass and primary productivity of birch (Betula pubescens s. lat.) in south-west Greenland. Journal of Ecology 62, pp. 821-830.

Farman, J.C. and Hamilton, R.A. 1978. Measurements of radiation at the Argentine Islands and Halley Bay 1963-1972. British Antarctic Survey Scientific Report No. 99.

Fliche, P. 1900. Note sur un bois fossile de Madagascar. Bulletin de la Société géologique de France 3° Serie 28, pp. 470-472.

Ford, E.D.A., Robards, A.W. and Piney, M.D. 1978. Influence of environmental factors on cell production and differentiation in the early wood of Picea sitchensis. Annals of Botany N.S. 42, pp. 683-692.

Frakes, L.A. 1979. Climates throughout geologic time. Elsevier, Amsterdam.

Francis, J.E. 1983. The dominant conifer of the Jurassic Purbeck formation, England. Palaeontology 26, pp. 277-294.

Francis, J.E. 1984. The seasonal environment of the Purbeck (Upper Jurassic) fossil forests. Palaeogeography, Palaeoclimatology, Palaeoclimatology 48, pp. 285-307.

Francis, J.E. 1987. (This volume). Palaeoclimatic significance of growth rings in Upper Jurassic and Early Cretaceous fossil wood from Southern England, pp. 21-36.

Frentzen, K. 1931. Die palaeogeographische Bedeutung des Auftretens von Zuwachszonen (Jahresringen) bei Holzern der Sammelgattung Dadoxylon Endl. aus dem Carbon und dem Rotliegenden des Oberrheingebietes. Zentralblatt für Mineralogie, Geologie und Paläontologie 1931B, pp. 617-624.

Fry, W.L. 1958. Petrified logs of Cupressinoxylon from the west shore of Chilko Lake, British Columbia. Bulletin of the Geological Survey of Canada 48, pp. 11-14.

Goldring, W. 1921. Annual rings of growth in Carboniferous wood. Botanical Gazette 72, pp. 326-330.

Gosz, J.R., Holmes, R.T., Likens, G.E. and Bormann, F.H. 1978. The flow of energy in a forest ecosystem. Scientific American 238(3), pp. 92-102.

Gothan, W. 1907. Die fossilen holzer von Konig Karls Land. Kungliga Svenska Vetenskapsakademiens Handligar 42, pp. 1-41.

Gothan, W. 1911. Die Jahresringlosigkeit der paläzoischen Baume und die Bedeutung dieser Erscheinung für die Beurteilung des Klimas dieser Perioden. Naturwissenschaftliche Wochenschrift 10, pp. 1-13.

Gregory, R.A. and Wilson, B.F. 1968. A comparison of cambial activity of white spruce in Alaska and New England. Canadian Journal of Botany 46, pp. 733-734.

Hallam, A. 1984. Distribution of fossil marine invertebrates in relation to climate. In: Fossils and Climate. (Ed. Brenchley, P.J.) Chichester: Wiley and Sons, pp. 107-125.

Hoeg, D.A. 1942. The Downtonian and Devonian Flora of Spitzbergen. Norges Svalbard-og Ishaves-Undersokelser 83, pp. 1-128.

Holden, R. 1913. Some fossil plants from eastern Canada. Annals of Botany 27, pp. 243-255.

Holden, R. 1917. On the anatomy of two palaeozoic stems from India. Annals of Botany 31, pp. 315-326.

Hoskins, J.T. and Cross, A.T. 1951. The structure and classification of four plants from the New Albany shale. American Midland Naturalist 46, pp. 684-716.

Hylander, C.J. 1922. A Mid-Devonian Callixylon. American Journal of Science 4, pp. 315-321.

Iurina, A. and Lemoigne, Y. 1972. Palaeoxylon Kazakstanensis: nouvelle structure ligneuse de type araucarien, du Dévonien supérieur du Kazakstan central (URSS). Compte rendu hebdomadaire des séances de l'Académie des Sciences de Paris D 274, pp. 814-817.

Jahnke, L.S. and Lawrence, D.B. 1965. Influence of photosynthetic crown structure on potential productivity of vegetation, based primarily on mathematical models. Ecology 46, pp. 319-326.

Jefferson, T.H. 1982. Fossil forests from the Lower Cretaceous of Alexander Island, Antarctica. Palaeontology 25, pp. 681-708.

Jefferson, T.H. 1983. Palaeoclimatic significance of some Mesozoic Antarctic fossils. In: Antarctic Earth Science (Proceedings of the 4th International Symposium on Antarctic Earth Sciences, Adelaide, 1982) (Eds. Oliver, R.L., James, P.R. and Jago, J.B.) Camberra: Australian Academy of Science, pp. 593-598.

Jensen, K.N. 1982. Growth rings in Pennsylvanian fossil wood found in Oklahoma. Oklahoma Geological Notes 42, pp. 7-10.

Khudayberdyev, R. 1962. Ginkgo wood from the Upper Cretaceous of south-west Kyzylkum. Doklady Academii nauk SSSR 145, pp. 422-424.

Kimura, T. 1984. Mesozoic floras of East and Southeast Asia, with a short note on the Cenozoic floras of Southeast Asia and China. Geology and Palaeontology of Southeast Asia 25, pp. 325-350.

Kozlowski, T.T. 1962. Photosynthesis, climate and tree growth. In: Tree growth (ed. Kozlowski, T.T.) New York: Ronald Press, pp. 149-164.

Kramer, P.J. 1943. Amount and duration of growth of various species of tree seedlings. Plant Physiology 18, pp. 239-251.

Kraus, G. 1974. Einige Bemerkungen uber Alter Wachstumverhaltnisse ostgronlandischer Holzgewachse. In: Die Zweite deutsche Nordpolarfahrt in den Jahren 1869 and 1870, vol. 2, 1, Leipzig.

Kräusel, R. and Jain, K.P. 1963. New fossil coniferous woods from the Rajmahal Hills, Bihar, India. Palaeobotanist 12, pp. 59-67.

Kräusel, R. and Range, P. 1928. Beiträge zur Kenntniss der karrooformation Deutsch-Sudwest Afrikas. Beitrage zur geologie Erforschung der Deutschen Schutzgebiete 20, pp. 1-55.

Kräusel, R. and Weyland, H. 1937. Pflanzenreste aus Devon X. Zwei Pflanzenfunde im Oberdevon der Eifel. Senckenbergiana Lethaea 19, pp. 338-355.

Lacey, W.S. 1953. Scottish Lower Carboniferous plants: Eristophyton waltoni sp. nov. and Endoxylon zonatum (Kidston) Scott from Dunbartonshire. Annals of Botany N.S. 17, pp. 579-596.

LaGrange, J.J. 1963. Trans-Antarctic Expedition 1955-58 Scientific Report no. 13. Meteorology 1. Shackleton, Southice and the journey across Antarctica. Trans-Antarctic Expedition Committee.

Lemoigne, Y., Iurina, A. and Snigirevskaya, N. 1983. Revision du genre Callixylon Zalessky 1911 (Archaeopteris) du Dévonien. Palaeontographica B 186, pp. 81-120.

Lemoigne, T. and Tyroff, H. 1967. Caractéres anatomiques d'un fragment de bois appartenant a l'espéce Walchia (Lebachia) piniformis du Permian d'Allemagne. Compte rendu hebdomadaire des séances de l'Académie des sciences 265, pp. 595-597.

Leo, R.F. and Barghoorn, E.S. 1976. Silicification of wood. Harvard Botanical Museum Leaflets, Harvard University 25, pp. 1-47.

Lepekhina, V.G. 1963. New discoveries of cordaitalean wood from the Upper Palaeozoic of Kazakhstan. Palaeontologicheskii Zhurnal 4, pp. 103-109.

Lepekhina, V.G. 1972. Woods of Palaeozoic pycnocylic gymnosperms with special reference to North Eurasia representatives. Palaeontographica B 138, pp. 44-106.

Long, A.G. 1979. Observations on the Lower Carboniferous genus Pitus Witham. Transactions of the Royal Society of Edinburgh 70, pp. 12-13.

Mädel, E. 1960. Monimaceen-Hölzer aus den oberkretazischen Umzamba-Schichten von Ost-Pondoland (S-Africa). Senckenbergiana Lethaea 41, pp. 331-391.

Maheshwari, H.K. 1972. Permian wood from Antarctica and revision of some Lower Gondwana wood taxa. Palaeontographica B 138, pp. 1-43.

Matten, L.C. and Banks, H.P. 1967. Relationship between the Devonian progymnosperm genera Sphenoxylon and Tetraxylopteris Bulletin of the Torrey Botanical Club 94, pp. 321-323.

Middendorf, A.T. von 1867. Die Gewächse Sibiriens. In: Reise in den aussersten Norden und Osten Sibiriens. 4, 1 St. Petersburg.

Monteith, J.L. 1973. Principals of environmental physics. London: Edward Arnold.

Negri, G. 1914. Sopra alenni legni fossili del Gebel Tripolitano. Bollettino della Società geologica italiana 33, pp. 321-344.

Nishida, M. 1973. On some petrified plants from the Cretaceous of Choshi, Chiba Prefecture VI. Botanical Magazine 86, pp. 189-202.

Oswald, H. 1969. Conditions forestiéres et potentialité de l'épicéa en haute Ardéche. Annales des sciences forestiéres 26, pp. 183-224.

Ovington, J.D. 1961. Some aspects of energy flow in plantations of Pinus sylvestris L. Annals of Botany N.S. 25, pp. 12-20.

Ovington, J.D. and Madgwick, H.A.I. 1969. The growth and composition of natural stands of birch. 1. Dry matter production. Plant and Soil 10, pp. 271-283.

Pole, M. 1982. The geology of Slope Point to Curio Bay. Unpublished Ms. Geology Department, Otago University, Otago, New Zealand.

Potenié, H. 1902. Fossile Hölzer aus der oberen Kreide Deutsch-Ostafrikas. in Die Reisen der Bergassessors Dr. Dantz in Deutsch-Ostafrika in den Jahren 1898-1900. Mitteilungen aus den deutschen Schutzgebieten Band 15 Heft 4, pp. 227-230.

Prasad, M.N.V. and Chandra, S. 1980. Palaeospiroxylon - A new gymnospermous wood from Raniganj coalfield, India. Palaeobotanist 26, pp. 230-236.

Ramunajam, C.G.K. and Stewart, W.N. 1969. Fossil woods of Taxodiaceae from the Edmonton Formation (Upper Cretaceous) of Alberta. Canadian Journal of Botany 47, pp. 115-124.

Read, C.G. 1932. Pinoxylon dakotense from the Cretaceous of the Black Hills. Botanical Gazette 93, pp. 173-189.

Roggeveen, P.M. 1932. Mesozoisches Koniferenholz (Protocupressinoxylon malayense n. sp.) ven der Insel Soegi in Riouw-Archipel, Niederland-isch Ost-Indien. Proceedings. Koninklijke nederlandse akademie wetenschappen, Amsterdam 35, pp. 580-584.

Russell, R.S. 1940. Physiological and ecological studies on an arctic vegetation. III. Observations on carbon assimilation, Carbohydrate storage and stomatal movement in relation to the growth of plants on Jan Mayen Island. Journal of Ecology 28, pp. 289-309.

Schultze-Motel, J. 1966. Gymnospermen-Hölzer aus den oberkretazischen Umzamba-Schichten von Ost-Pondoland (S-Africa). Senckenbergiana Lethaea 47, pp. 279-337.

Schweitzer, H.-J. 1968. Die flora des oberen Perms in Mitteleuropa. Naturwissenschaftliche Rundschau 21, 93-102.

Seward, A.C. and Walton, J. 1923. On fossil plants from the Falkland Islands. Quarterly Journal of the Geological Society 79, pp. 313-333.

Shilkina, I.A. 1960. Cordaitalean wood - Taxopitys arctica sp. nov. from the Upper Carboniferous of Eastern Siberia. Palaeontologicheskii Zhurnal 3, pp. 123-126.

Shilkina, I.A. 1963. A new conifer genus Yatsenkoxylon sibiricum gen. et. sp. nov. Doklady Academii nauk SSSR 148, pp. 163-165.

Shimakura, M. 1936. Studies on fossil woods from Japan and adjacent lands. Contribution I. Some Jurassic woods from Japan, Saghalien and Manchoukuo. Science Reports of Tohoku Imperial University, 2nd Series 18, pp. 267-310.

Shimakura, M. 1937. Studies on fossil woods from Japan and adjacent lands. Contribution II. The Cretaceous woods from Japan, Saghalien and Manchoukuo. Science Reports of Tohoku Imperial University, 2nd Series 19, pp. 1-73.

Sitholey, R.V. 1940. Jurassic plants from Afghan-Turkistan. Memoirs of the Geological Service of India, Palaeontologia Indica 29 Memoir No. 1.

Smith, A.G., Hurley, A.M. and Briden, J.C. 1981. Phanerozoic paleocontinental world maps. Cambridge: University Press.

Stopes, M.C. 1916. An early type of the Abietineae(?) from the Cretaceous of New Zealand. Annals of Botany 30, pp. 111-125.

Tarling, D.H. 1983. Palaeomagnetism. London: Chapman and Hall.

Tieszen, L.L. 1978. Summary. In: Vegetation and Production Ecology of an Alaskan arctic tundra. (Ed. Tieszen, L.L.) New York, Heidelberg: Springer-Verlag, pp. 622-645.

Tynan, E.J. 1959. Occurrence of Cordaites michiganensis in Oklahoma. Oklahoma Geological Notes 19, pp. 43-46.

Unger, F. 1959. Der versteinerte Wald bei Cairo und einige andere Lager verkieselten Holzen in Agypten. Sitzungsberichte der Akademie der Wissenschaften in Wien 33, pp. 209-233.

Vaartaja, O. 1962. Ecotypic variation in photoperiodism of trees with special reference to Pinus resinosa and Thuja occidentalis. Canadian Journal of Botany 40, pp. 849-856.

Warren Wilson, J. 1964. Annual growth of _Salix arctica_ in the high-arctic. _Annals of Botany_ N.S. 28, pp. 71-76.

Warren Wilson, J. 1966. An analysis of plant growth and its control in arctic environments. _Annals of Botany_ N.S. 30, pp. 383-402.

Wassink, E.C. and Wiersma, J.H. 1955. Daylength responses of some forest trees. _Acta botanica neerlandica_ 4, pp. 657-670.

Williams, S. 1930. The geological collection from the South Central Sahara made by Mr. F. Rodd. III. Fossil wood. _Quarterly Journal of the Geological Society London_ 86, pp. 408-409.

Wilson, L.R. 1963. A new species of _Dadoxylon_ from the Seminole Formation (Pennsylvanian) of Oklahoma. _Oklahoma Geological Notes_ 23, pp. 215-220.

Zalessky, M. 1909. Communication preliminaire sur un nouveau _Dadoxylon_ provenant du Dévonien supérieur du bassin du Donetz. _Bulletin d'l'Académie imperiale des Sciences de St. Pétersbourg_ 26, pp. 1175-1178.

Zalessky, M. 1911. Etude sur l'anatomie due _Dadoxylon tchihatcheffi_ Geoppert. _Mémoires du comité géologique Russe de St. Pétersbourg_ 68, pp. 18-29.

Relationships between British Climate and the Radial Growth of Quercus Species - an Empirical Approach to Climate Reconstruction

K.R. Briffa

Climatic Research Unit
School of Environmental Sciences
University of East Anglia
Norwich NR1 7TJ

ABSTRACT

This paper illustrates some of the results of recent dendroclimatic reconstruction work at the Climatic Research Unit. The work described here is restricted solely to the use of oak ring width chronologies, by which is meant Quercus petraea (Mattuschka) Liebl., Q. robur L. and their hybrids.

Examples are given of simple and multiple regressions on climate variables representative of small and larger regional scale averaging. Preliminary results are also shown of direct spatial reconstruction of patterns of summer mean sea-level pressure over Great Britain. Emphasis is placed on the need to assess the quality of empirically based reconstructions by reference to actual climate data for a period independent of that used for calibrating regression models.

INTRODUCTION

On the millenium and shorter time scales, a knowledge and understanding of climate trends and variability is essential for a correct appreciation of the context of present climate and also for framing our most immediate climate expectations. The conception, development, testing and refinement of climate models is critically dependent on the availability and quality of observational data. Climate data come from instrumental records of meteorological phenomena and from evidence of climate conditions inferred indirectly from various, so called 'proxy-data'.

Large scale networks of continuous monthly averaged data on temperature and precipitation only extend back to about 1850. Even after 1850, the spatial distribution of the data is heavily biased towards Europe and North America. Improvements to the land-based temperature and precipitation records continue to be made, though the data remain concentrated in the Northern Hemisphere (Bradley et al., 1985). Gridded monthly mean sea-level pressure data extend back to the 1870s for parts of the Northern Hemisphere, though even these are not without data quality problems (Williams and van Loon, 1976; Trenberth and Paolino, 1980). Other data bases, such as upper air data and satellite data, are comparatively short (MacCracken and Luther, 1985).

Faced with this spatially and temporarily uneven distribution of data, palaeoclimatologists have turned to indirect records to expand and extend their knowledge of past environmental conditions. Among the various

Figure 1. A selection of oak chronologies in the British Isles and northern France which have been used in the work described here. The sites are: 1, Radley (Briffa and Jones, unpublished data); 2, Charlbury (Briffa and Jones, unpublished data); 3, Yanworth (Briffa and Jones, unpublished data); 4, Monk Wood (Briffa and Jones, unpublished data); 5, Hesleyside Hall (Briffa and Jones, unpublished data); 6, Bocconoc (Briffa and Jones, unpublished data); 7, Clovelly (Briffa and Jones, unpublished data); 8, Forest of Dean (Briffa and Jones, unpublished data); 9, Sotterley Park (Briffa and Jones, unpublished data); 10, Rostrevor (Pilcher, 1976); 11, Raehills (Pilcher and Baillie, 1980b); 12, Ludlow (Pilcher and Baillie, 1980b); 13, Lough Doon (Pilcher and Baillie, 1980a); 14, Ardara (Pilcher and Baillie, 1980a); 15, Oxford (Pilcher and Baillie, 1980b); 16, Killarney (Pilcher and Baillie, 1980a); 17, Blickling (Pilcher and Baillie, 1980b); 18, Bath (Pilcher and Baillie, 1980b); 19, Scorton (Pilcher and Baillie, 1980b); 20, Lockwood (Pilcher and Baillie, 1980b); 21, Enniscorthy (Pilcher and Baillie, 1980a); 22, Cappoquin (Pilcher and Baillie, 1980a); 23, Glen of the Downs (Pilcher and Baillie, 1980a); 24, Fontainebleau (Pilcher, unpublished data); 25, Chinon (Pilcher, unpublished data); 26, Chambord (Pilcher, unpublished data); 27, Halate (Pilcher, unpublished data); 28, Guines (Pilcher, unpublished data); 29, Glen Luce (Pilcher and Baillie, 1980b); 30, Belfast* (Baillie, 1973, 1977); 31, Maentwrog (Leggett et al., 1978, Hughes et al., 1978a); 32, Peckfortan (Leggett, unpublished data); 33, Cannock (Hughes, unpublished data); 34, N.E. England (Castle Howard) (Morgan, unpublished data); 35, Salisbury Plain (Barefoot, 1975); 36, Hursley Park (Barefoot et al., 1974)

* Regional chronology

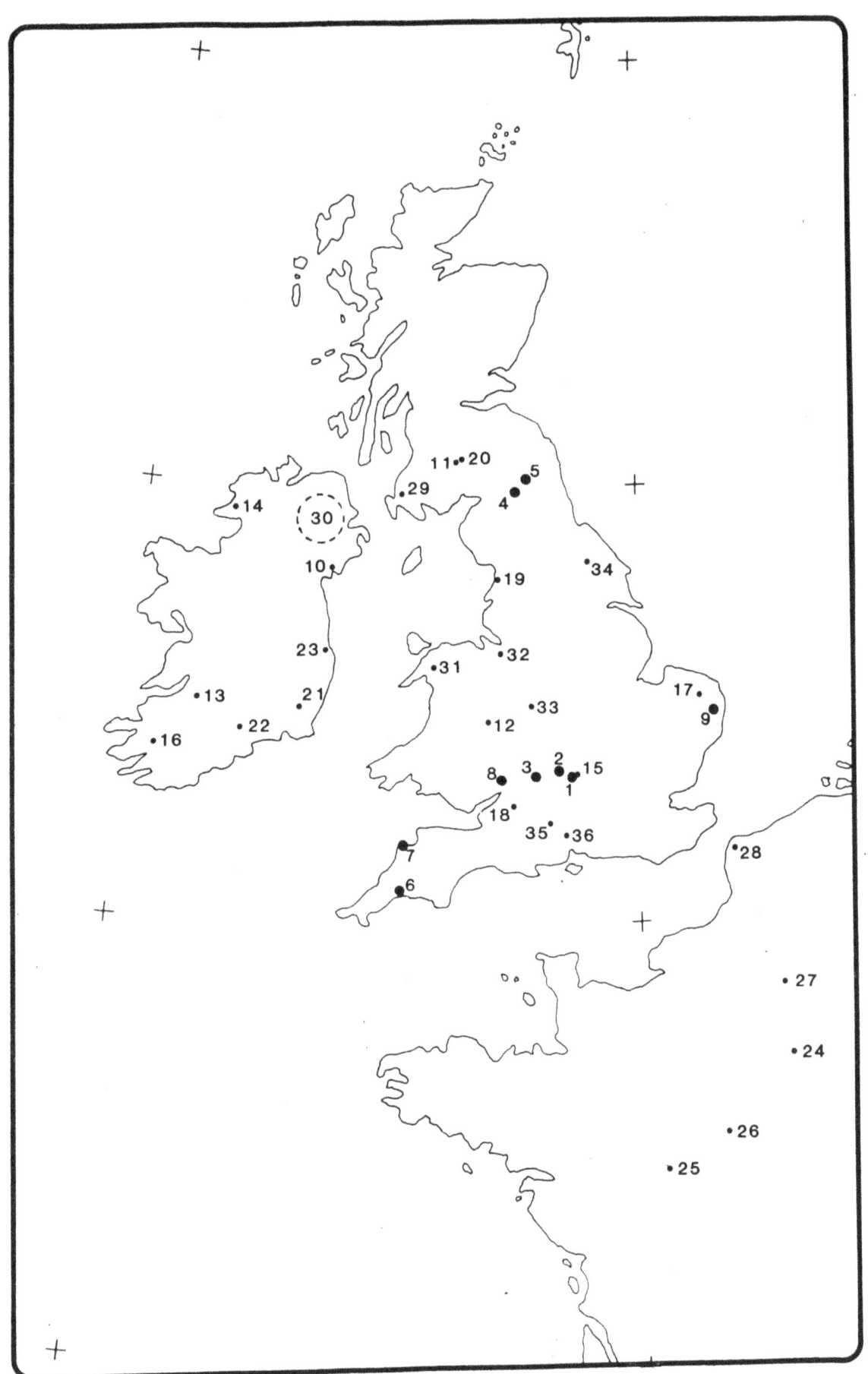

sources of proxy climate information, tree rings are becoming increasingly important.

The annual growth of a tree is the net result of many complex and interrelated biochemical processes (Kozlowski, 1962; Kramer and Kozlowski, 1962; Zimmerman, 1964; and Fritts, 1976). Trees interact with the micro-environments of the leaf and root surfaces. The relationship of these extremely localised conditions to larger scale climate parameters offers the potential for estimating the overall influence of large scale climate on growth from year to year. Wherever tree growth is limited directly or indirectly by some climate variable, and that limitation can be quantified and reliably dated, dendroclimatology can be used to reconstruct some information about past environmental conditions.

There are several sub-fields of dendroclimatology associated with the processing and interpretation of different tree growth variables (Creber, 1977; LaMarche, 1978). These include the analysis of various densitometric parameters (Schweingruber et al., 1978, 1979; Hughes, this volume) and the analysis of chemical or isotopic variables (Jacoby, 1980; Gray, 1981; Long, 1982; Wigley, 1982). The most easily exploited source of information, however, lies in the yearly variations of the radial growth of trees, as measured in the width of their annual rings. Ring-width chronologies can provide high resolution data from large areas of the globe which is continuous, relatively long, absolutely dated, and well-replicated. They are relatively simple and inexpensive to produce. For these reasons they are a uniquely valuable source of proxy-climate information.

THE CHRONOLOGY DATA

Figure 1 shows many of the oak chronologies in the British Isles and northern France produced at the time of writing. This is by no means comprehensive, but includes most of the data available to the Climatic Research Unit when the work described here was being carried out. The network has recently been expanded and further chronologies are currently under construction.

In addition to the effects of climate, the measured ring-width series of many European oaks reflect inherent aging effects and the influences of competition and management. These can combine to produce an ambiguous mixture of relatively low and medium frequency variations spanning timescales of the order of several centuries to several decades. Simply averaging the raw ring-width data from a number of trees does not efficiently remove these non-climatic variations. Some form of standardisation is necessary: i.e. the removal of relatively low frequency variance from individual ring-width series prior to averaging to form a chronology (Fritts, 1976). The data here have been standardised by dividing the raw data from each series by the equivalent annual values from a curve of the same data smoothed with a Gaussian filter. In an analysis of data from 36 sites it was found that a filter designed to pass greater than 50 per cent of variations with amplitudes greater than 30 years is optimal for maximising the statistical quality of the data as a whole, while minimising the potential loss of low and medium frequency climate information (Briffa, 1984). Such a 30-year filter was used in the work described here.

SIMPLE REGRESSION

A group of five site chronologies enclosing an area of about 3,500 square kilometres surrounding the catchment of the upper River Thames can be used to illustrate single site and relatively small regional tree/climate relationships. These chronologies are numbers 1, 2, 3, 15 and 18 in Figure 1. Correlations between the chronologies (1838-1977) range from 0.27 to 0.65, those with the Yanworth site (no. 3) being smallest. These chronologies show little if any relationship with monthly mean temperatures at Oxford. Table 1 shows the correlation coefficients between the five chronologies and various monthly precipitation totals at Oxford (precipitation data from Craddock and Craddock, 1977 and Craddock and Smith, 1978). Correlations are also shown with two regional chronologies formed by averaging first, all five site chronologies and secondly, all chronologies except for the site at Yanworth. A number of variables representing various precipitation totals are also correlated with these individual site and composite regional chronologies.

The individual monthly correlations are significant for several spring and summer months for all chronologies with the notable exception of the Yanworth series. Interestingly, two sites show a relationship with April rainfall significant at the 0.01 level, whereas two have little relationship for this month. Almost all the correlations for the sites other than Yanworth are significant at better than the 0.05 level for May, June and July, with the July values most significant. By August there is no relationship between ring widths and rainfall at any site.

The various seasonal rainfall parameters show high correlations with individual site chronologies (other than Yanworth) with April-July or May-July generally the largest. The relationships between these seasonal parameters are increased by moving to regionally-averaged chronologies. Not unexpectedly, the four-site average shows the strongest relationship again with little to choose between the April-July and May-July values. Though they account for only 25 per cent of the variance, these correlations are significant at the 0.001 level.

The temporal stability of these correlations can be assessed by calculating values for the two sub-periods 1838 to 1907 and 1908 to 1977. These are shown in Table 2. Many correlations associated with the individual site chronologies, though relatively high (e.g. 0.55 for the later period for April-August with Radley) are not stable over both periods. Those correlations involving the four-site regional chronology are, however, both high and stable.

Soil analyses of four of the sites (the Bath site was not analysed) shows Charlbury to be a typical Cotswold brash soil of the Sherborne series (Robinson, 1948), shallow and well drained. Trees on such soils would be expected to experience significant water shortage in summer months. Another site, Otmoor (part of Baillie and Pilcher's 'Oxford' site) is, in complete contrast, a typical pelo-stagnogley soil of the Denchworth series (Kay, 1934). Such surface water gley soils, while subject to frequent waterlogging in winter, do dry out in summer and the shallow rooting of the trees at this site suggests that they are subject to drastic changes in available soil water associated with rapid movements in the height of the water table. Both the Yanworth and Radley sites are on soils of the Evesham series (Osmond et al., 1949; Jarvis, 1973). These are relatively deep, clay soils and poorly drained. Though not prone to prolonged

Table 1. Correlation coefficients (1838-1977) for monthly and seasonal Oxford rainfall totals correlated with each of the five local series and two average chronologies described in the text. The four-site average excludes the Yanworth chronology.

	RADLEY	CHARLBURY	OXFORD△	BATH	YANWORTH	AVERAGE[1]	AVERAGE[2]
January	0.13	0.16▲	0.20▲	0.04	0.22•	0.20▲	0.16▲
February	0.20▲	0.12	0.16▲	0.01	0.00	0.13	0.15
March	-0.03	0.00	0.04	-0.03	0.04	0.00	-0.01
April	0.21•	0.09	0.22•	0.08	0.11	0.19▲	0.19▲
May	0.27•	0.16▲	0.21•	0.18▲	0.09	0.24•	0.25•
June	0.20▲	0.14	0.19▲	0.25•	-0.07	0.19▲	0.24•
July	0.35•	0.36•	0.30•	0.35•	0.00	0.36•	0.42•
August	0.10	-0.02	0.08	0.12	-0.08	0.06	0.09
September	0.06	0.01	0.00	-0.12	-0.09	-0.04	-0.02
Annual	0.06	-0.04	0.02	0.02	0.02	0.02	0.01
April-August	0.46•	0.30•	0.40•	0.41•	0.01	0.42•	0.49•
April-July	0.48•	0.36•	0.43•	0.42•	0.05	0.46•	0.52•
April-June	0.38•	0.22•	0.35•	0.30•	0.06	0.35•	0.39•
May-August	0.42•	0.30•	0.36•	0.41•	-0.03	0.39•	0.46•
May-July	0.45•	0.37•	0.39•	0.43•	0.01	0.44•	0.51•
March-May	0.25•	0.14▲	0.26•	0.12	0.13	0.24•	0.24•
June-August	0.34•	0.25•	0.30•	0.37•	-0.07	0.32•	0.39•

Table 2. Correlations for Early (1838-1907) and Late (1908-1977) periods for five Oxfordshire chronologies (see Figure 1) and two averages of these (the AVERAGE 2 chronology excluding Yanworth). The dependent variables are seasonal totals of Oxford rainfall.

		RADLEY	CHARLBURY	OXFORD△	BATH	YANWORTH	AVERAGE[1]	AVERAGE[2]
April-August	E	0.38•	0.28▲	0.43•	0.43•	-0.05	0.43•	0.52•
	L	0.55•	0.33•	0.40•	0.39•	0.07	0.44•	0.48•
April-July	E	0.41•	0.24▲	0.42•	0.41•	-0.09	0.41•	0.51•
	L	0.58•	0.48•	0.45•	0.44•	0.19	0.53•	0.56•
April-June	E	0.37•	0.12	0.39•	0.30•	-0.01	0.34•	0.40•
	L	0.42•	0.32•	0.33•	0.31•	0.14	0.38•	0.40•
May-August	E	0.31•	0.30•	0.41•	0.45•	-0.08	0.40•	0.50•
	L	0.53•	0.30•	0.33•	0.38•	0.02	0.40•	0.45•
May-July	E	0.34•	0.25▲	0.40•	0.43•	-0.12	0.38•	0.49•
	L	0.58•	0.47•	0.40•	0.45•	0.14	0.51•	0.55•
June-August	E	0.26▲	0.29▲	0.33•	0.40•	-0.13	0.31•	0.42•
	L	0.48•	0.23▲	0.29•	0.35•	-0.02	0.34•	0.39•
Annual	E	0.19	0.01	0.06	0.11	0.16	0.15	0.12
	L	-0.11	-0.11	-0.02	-0.01	-0.11	-0.09	-0.07

• Significant at the 0.01 level

▲ Significant at the 0.05 level

△ This chronology is number 15 in Figure 1.

waterlogging they can be slow to dry in spring. However, there are differences in aspect, elevation and, especially, soil depth between these last two sites. The Yanworth site is situated on a hillside on a gentle slope facing to the south and southeast. Here, the soil is extremely deep and the tree roots penetrate a long way down. In contrast, the Radley site is in a complex area where underlying patches of Terrace gravels can produce patches of loamy argillic brown earths of the Sutton series (M.G. Jarvis - personal communication) and some gravel is apparent at the one pit dug at this site. Generally this soil is better drained than the Yanworth soil.

Despite their having a mix of soil conditions, four of the site chronologies show a similar relationship with rainfall in summer months. The Yanworth site, though not from any external examination obviously different from the others, is clearly anomalous. It is probable that rainfall on timescales of several months is effectively smoothed by the integrating effect of slow water movement at this site. Significant lateral waterflow was seen during soil analysis at the site and the long residence time for soil water could explain the lack of dependence on summer rain.

The simple correlation analyses described above show how climate-tree-ring links can be strengthened by selectively incorporating the information from individual chronologies into a regional one, while empirically identifying the optimum season to which it relates. The practicability and general applicability of this approach is, of course, highly dependent on being able to judge how much spatial averaging is appropriate. This cannot be done on the basis of simple distance criteria. There must be both upper and lower limits. The high spatial variability of summer rainfall in Britain, even over small distances (Tabony, 1977; Wigley et al., 1984a), must limit the geographical range over which it is sensible to average chronologies.

Simple averaging of chronologies may not be the best method of using the tree-ring data because different sites may vary noticeably in their responses to climate (as indicated in the above analysis). It is possible that trees growing in a particular region may contain some unique climate information. Averaging them, although shown here to improve response stability, nevertheless, affords little control over what information is maintained and what is lost. Though common variations may be emphasised, potentially useful, but less generally represented variance may be swamped. An alternative approach, which offers the possibility of using some of the less commonly represented variations within the individual site chronologies, is that of multiple regression.

MULTIPLE REGRESSION

The Method

In an attempt to overcome statistical instability which is inherent in multiple regression involving intercorrelated variables, a principal component regression approach has been developed. The details of the method as applied to the reconstruction of single variable climate series are described in Briffa et al., (1983) and Briffa (1984). Basically, the intercorrelation between the ring-width variables ensures that much of the variance between them is associated with a few low order principal components, and objective criteria, defined a priori, may be used to discard many of the high order components. The variance thus removed is indistinguishable from statistical 'noise'. The remaining 'candidate' predictors are then regressed singly against the predictand. A stepwise technique is not necessary because of the orthogonality of the component predictors. If desired, a further reduction in the number of predictors can be made at this stage: for example, by retaining only those candidate predictors with t-values above some predetermined level. The resulting equation which is in terms of a small number of final principal component predictors, is then transformed back to one which is in terms of all the original chronology variables.

This procedure is carried out over what is termed the <u>calibration period</u> and produces a <u>transfer function</u>, i.e. a set of regression weights, one for each of the original chronology variables. This transfer function can then be applied to independent data to produce a reconstruction for any period for which all the predictor (tree-ring) data exist.

Models

The simplest multiple regression model involves the prediction of climate in the current year (i) through a transfer function involving only ring widths for year i from each of the predictor chronologies.

This can be expressed as

$$C_i = f(W_{ij}) \qquad \qquad \ldots\ldots (1)$$

where C_i is the climate in year i, and

W_{ij} are the ring widths at the various sites (j=1, 2...) for year i.

More complex models can involve the use of predictors from years immediately leading or lagging (or both) the climate year being reconstructed. This is the approach commonly adopted with arid site conifers. These chronologies are generally highly autocorrelated even for lags of up to 10 years. Fritts (1976, p. 234) has discussed physiological mechanisms whereby the climate of one year can affect conifer growth in future years. Nevertheless to date, reconstructions based on such series generally involve empirical modelling of combinations of lagged yearly ring width predictors (Fritts <u>et al.</u>, 1979).

Despite the lack of understanding of growth preconditioning in deciduous hardwood species, the autocorrelation functions of oak ring-width series do provide some statistical basis for experimenting with <u>simple</u> multiple year models. Oak chronologies can be quite highly autocorrelated, but rarely for lags of more than one or two years. On this evidence only simple lagged-year models are justified. Examples are (deleting subscript j for simplicity)

$$C_i = f(W_i-1, W_i) \qquad \qquad \ldots\ldots (2)$$

$$C_i = f(W_i, W_{i+1}) \qquad \qquad \ldots\ldots (3) \text{ and}$$

$$C_i = f(W_{i-1}, W_i, W_{i+1}). \qquad \qquad \ldots\ldots (4)$$

The use of all of these has been explored.

As in any regression exercise, the larger the number of predictors used, the greater will be the calibrated variance. The statistical significance of a calibration relationship based on principal component regression is the source of some uncertainty. This is espeically so if some noise reduction has been used and if an additional screening and selection of predictors is made (see Wigley and Tu, 1983, and Briffa et al., 1983 for further discussion). Model reliability can, however, be more directly assessed on the basis of an independent period validation or verification. This simply involves the comparison of the actual and reconstructed climate values over a period which was not used to calibrate the model. A number of tests are generally used (Fritts, 1976). For simplicity the discussion here refers only to simple correlation coefficients.

Relatively small regional climate reconstruction

Current year ring widths from all five of the 'Oxfordshire' site chronologies were used to reconstruct April-July Oxford rainfall totals. The regression was first calibrated over 1838 to 1907 and verified over the period 1908 to 1977 (early calibration) and then the periods were reversed (late calibration). The early calibration multiple regression value was 0.57 with a verification period simple regression coefficient of 0.51. The corresponding late calibration results were 0.56 and 0.53. These results are marginally better (though not significantly so) than those achieved with the best of the simple regression results, which related precipitation to the simple average of a number of chronologies.

Larger-area-averaged climate reconstruction

The multiple regression method becomes more valuable in reconstructing larger-scale average climate variables. This is shown in reconstructions of summer England and Wales rainfall (Nicholas and Glasspoole, 1931, and updated by the U.K. Met. Office), and Central England temperature (Manley, 1974). Reconstructions of April to August rainfall and June to August temperature have been made using a 14-site network of chronologies (Briffa et al., 1983). The individual chronologies show little correlation with these variables and only occasionally are correlation coefficients similar when calculated over different 70-year periods (the Fontainebleu series is an exception correlating with the rainfall at 0.41 over both periods). A chronology formed by averaging all of the series is also poorly correlated with both the climate variables over both periods. The impracticability of simply averaging chronologies over these distances is emphasised by these results but the possibility of smaller-scale coherence in climate response should dictate a judicious approach to chronology averaging.

Multiple regression models incorporating as predictors only ring widths from the year corresponding to the predictand climate data (Equ. 1) produced multiple correlations of 0.63 in calibration (1830-1899) and 0.41 in verification (1900-1969) for rainfall. The corresponding temperature results were 0.58 and 0.48. Reversing the calibration/verification periods gave rainfall results of 0.65 in calibration and 0.52 in verification. The equivalent temperature results were 0.44 and 0.50 respectively.

Central England Temperature (June to August average)

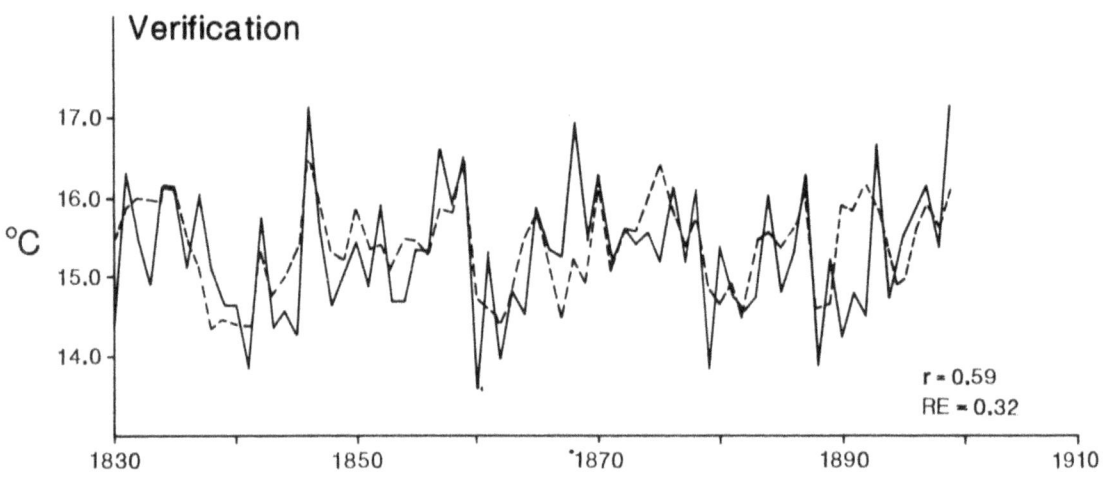

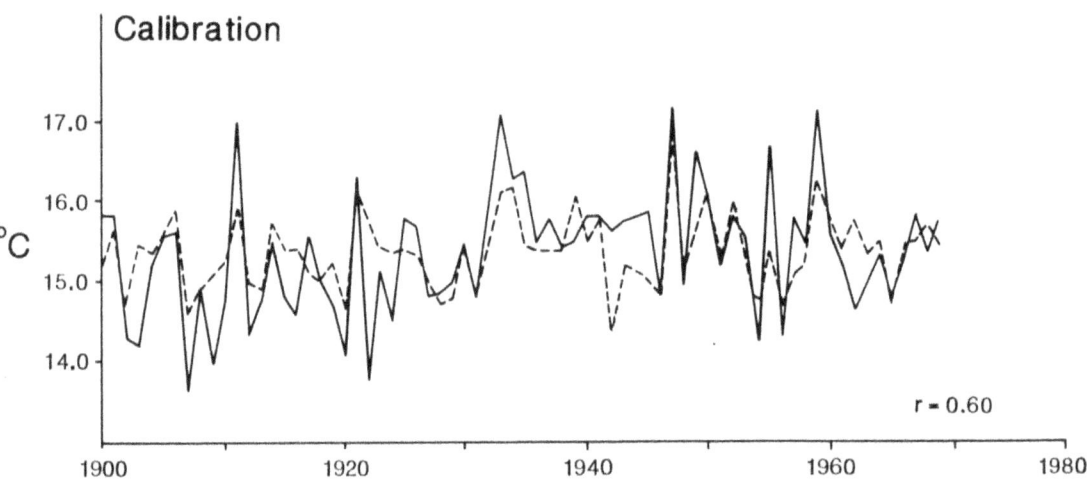

Figure 2. Actual (solid line) and reconstructed (dotted line) summer Central England temperatures based on a 21-chronology network.

Incorporating previous year-ring width variables as predictors (Equ. 2) necessarily improves the calibrated variance, but did not produce any systematic improvement in the results over the verification period. Subsequent work with a larger (21-site) network has produced better results for the temperature reconstruction.

An example reconstruction, for June-August mean Central England temperature is shown in Figure 2. This is based on the simple, current year model (Equ. 1) with ring widths from 21 sites. The calibration multiple correlation coefficient (1900-1969) was 0.60, and the verification simple correlation coefficient was 0.59 (1830-1899). The largest regression weights in the reconstruction equation are for the sites at Lockwood in Scotland and Castle Howard in Yorkshire (positive) and Chinon in France (negative). The growth at the first two sites is known from other analysis to respond predominantly and positively to growing season temperatures. At Chinon, growth is mainly affected by rainfall (Briffa 1984). In general, chronologies which show simple, strong relationships with the dependant climate variables in a correlation analysis tend to contribute most of the weight in multiple regression equations, but this is by no means always the case.

Multiple regression tends to produce results which are better than those obtained either by averaging the chronologies or by using simple regression. Since multiple orthogonalised regression is of general applicability and since it is <u>at least</u> as good as simple regression, it is the preferred technique; but simpler analysis techniques should not be rejected.

Using non-linear climate predictands

i) Cumulative Effective Rainfall

Plotting the actual against reconstructed summer rainfall values at Oxford for both the calibration and verification periods shows that the residual variance is mostly associated with high rainfall events. The reconstruction badly underestimates these. This suggests that better results could be achieved by reconstructing some variable representing degrees of water shortage.

One such series is cumulative effective rainfall (CER). This is derived from a series of effective rainfall (ER) given by

$$ER_i = R_i - E_i \qquad \ldots(5)$$

where E_i is the evaporation and R_i is the measured rainfall in month i.

Monthly 'Oxfordshire' effective rainfall values were calculated using Oxford rainfall and constant monthly E values computed by Wright (1978) for the upper Thames catchment above Eynsham. From these a CER series was calculated using

$$CER_i = CER_i - 1 + ER_i \qquad \ldots(6)$$

Where CER_i is the cumulative effective rainfall to month i. Note that positive values are set to zero so that the final series consists of zero and negative values. Generally, values between November and March are zero.

The simple correlations between the four-site regional Oxfordshire chronology and the CER series average for April to July is -0.55 over the period 1838-1907 and -0.62 for 1908-1977. The multiple correlation method gives slightly higher values in calibration but verifies with similar values.

ii) Riverflow

An example of a natural integration of seasonal temperature and rainfall is provided by river flow data. Riverflow is largely the result of a balance between precipitation and evaporation over a catchment. As temperature and rainfall both influence tree growth and river flow in complex ways, the linking of riverflow to tree growth, known as dendrohydrology, is a logical extension of dendroclimatology. Though more generally associated with arid environments (e.g. Stockton, 1975; Smith and Stockton, 1981), hydrological reconstructions using tree rings have been produced in other areas (e.g. Holmes et al., 1979; Cook and Jacoby, 1983).

The principal component regression method described above has been used recently with a predictor network of seven oak chronologies in southern Britain and northern France to reconstruct riverflow for three catchments in southern Britain (Jones et al., 1984). The results, the first such reconstruction in Europe, were shown to be generally more reliable for low flow events. Such data may be of potential value for water resource planning because of the relatively short time series available for measured flow.

Spatial Climate Reconstruction

The principal component multiple regression method has been further developed to allow the direct reconstructions of spatial patterns of climate.

This is accomplished by first decomposing the response or predictand data into their principal components and developing separate regression equations for each amplitude series deemed to be significantly distinguishable from statistical noise. The regression equations, each derived using the same method described for the single-variable work, are then combined and transformed back to give a regression equation for each point of the predictand grid. Different aspects of the methodology and mathematical details of this process are given in Wigley and Tu (1983), Jones et al., (1983), Briffa (1984) and Briffa et al., (1986). The last two of these works describe the results of one reconstruction of mean sea-level pressure over Great Britain. A brief account is given here.

A 16-point pressure grid from 45° to 60°N and 20°W to 10°E was reconstructed. The simplest model was used. This related pressure for year i to a network of ring widths in the same year at 14 sites (numbers 3,4,5,6,7,8,9,17,18,20,24,26,31 and 33 in Figure 1). The 102 years of overlap between pressure and chronology data was divided, and the period 1873 to 1922 used to calibrate the model and 1923 to 1974 kept for verification. Six principal components of the pressure set were deemed to be significant. The three most important accounted for 51.4, 24.2 and 12.7 per cent of the total variance respectively. The loadings of the

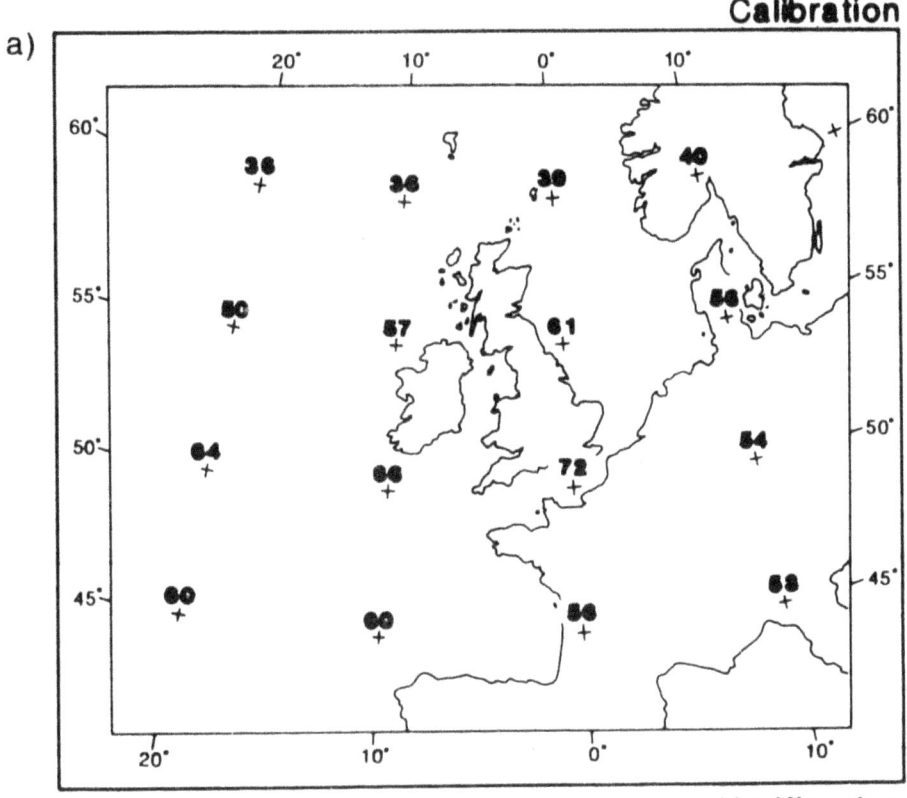
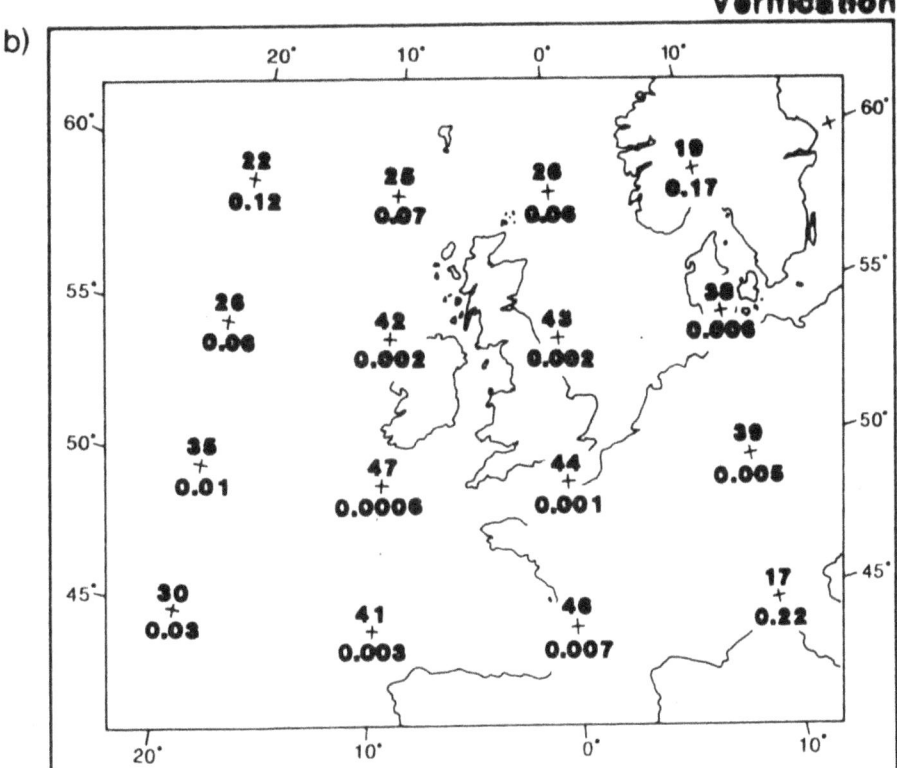

Figure 3. (a) the multiple correlation coefficients (x 100) for the reconstruction of May-July mean sea-level pressure (1873-1922) made with a 14-site chronology network.
(b) the equivalent verification values (upper figures) over the independent period (1923-1974) and the statistical significance levels (lower figures) for these values.

first component were all of the same sign forming a concentric pattern with largest weights in the centre of the grid and diminishing weights towards the edges. The second and third components each have patterns with a diagonal band of near zero weights with weights of opposite sign on either side. The second component has large negative loadings in the northwest near Iceland and large positive weights in the southeast over northern Italy. The third component pattern has large negative weights in the northeast over southern Sweden and positive weights in the southwest. The final predictors were chosen from among the principal components of the ring-width data by accepting only those with an absolute t-value (for the simple correlation coefficient with the dependent pressure component) of 1.0 or greater. On this basis none of the variance of the fourth pressure amplitude series was calibrated, but the multiple correlation coefficients for the other five ranged from 0.65 (for the first pressure component) to 0.20 (for the fifth pressure component). Verification period correlation coefficients showed that only the first two pressure components were reasonably reconstructed. Over the whole 52-year independent period these were 0.47 and 0.40 for the first and second components respectively.

When the results for all 16 grid points are averaged, 30 per cent of the total variance is shown to be explained in the calibration period and just over 12 per cent in the verification period. These figures disguise the fact that the reconstructions are significantly better in the central area of the grid. This is shown in figure 3. The results are poor around the periphery of the pressure grid, but this might be expected given that the predictor chronologies are all situated within or very near to the four central grid points. Over the independent verification period, the reconstructions for these four points are all highly statistically significant (at the 0.0002 level or better). An example plot of the actual and reconstructed May to July mean sea-level pressure for the grid point 50°N, 10°W is presented in Figure 4.

The regression weights for the reconstruction shown in Figure 4 are largest on the chronologies at Bath, Sotterley Park in Suffolk, Boconnoc in Cornwall and Fontainbleau near Paris. Monitoring the statistical quality of these chronologies can highlight any potential error that arises as a function of decreasing core numbers in early sections of chronologies. This is additional to any error inherent in the calibrated model (Wigley et al., 1984b). Of the above four chronologies, all except Bath are of acceptable quality at least back beyond 1830. Bath, however is somewhat suspect before about 1850. As this is the most important predictor, any further work that could increase its length and particularly its early replication would be beneficial.

Conversely, for any one chronology the relevant regression weights (one from each of the pressure grid point equations) can be mapped to show the influence of that chronology in the different areas of the pressure grid. An example is given in Figure 5, which shows the normalized regression weights on the sixteen pressure grid points for the Scottish chronology, Lockwood.

A comparison of the actual and reconstructed anomaly maps, year by year, reveals that poorly reconstructed summers often correspond to near normal observed pressure over Great Britain with strong anomalies centred near the edge of the grid, e.g. over Scandinavia or central Germany. The reconstructed maps in these instances often show a lack of any strong pressure gradients over the map. In contrast, years reconstructed as

Grid point 50°N 10°S

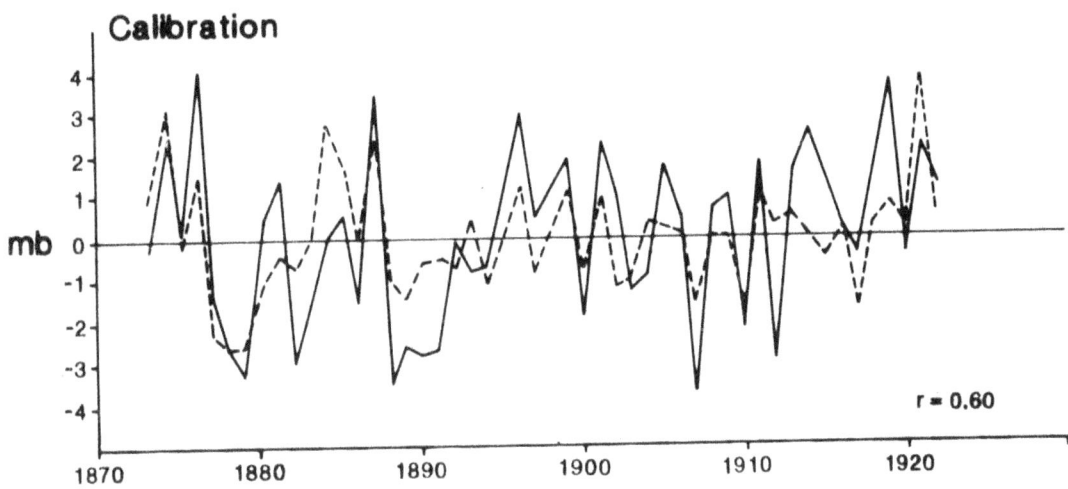

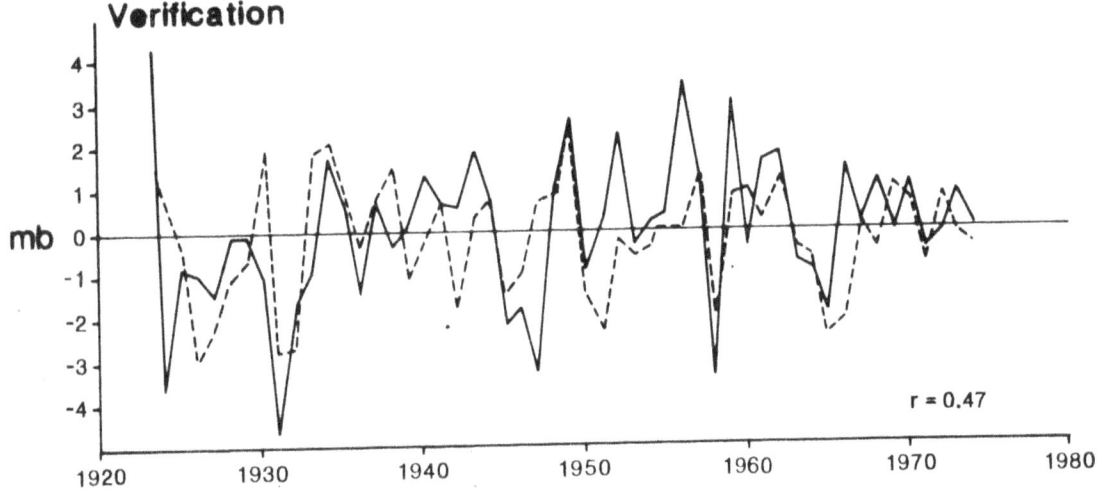

Figure 4. The observed (solid line) and reconstructed (dotted line) sea-level pressure anomalies for grid-point 50°N 10°W produced in the reconstruction described in the text.

having relatively strong or well defined patterns are generally more accurate.

Expanding the chronology network, if possible beyond the extent of the reconstruction area, should significantly improve the reconstructions by enabling the predictor set to register more easily pressure changes beyond the presently limited area in the centre of the grid.

As the commencement of the gridded pressure data is 1873 and the earliest year common to all of the 14 chronologies used here is 1755, it is possible to produce reconstructions back to that time. However, given that the statistical quality of the early tree-ring data is poor, and considering the poor results for the grid as a whole, little confidence could be placed in such reconstructions in this preliminary work. However, an interesting reconstruction example is presented in Figure 6, May to July mean sea-level pressure for 1816. This was chosen because historical evidence suggests that this was a particularly poor summer in western Europe, perhaps associated with the eruption of Mount Tambora in April 1815 (Stommel and Stommel, 1979). This year has become known as the 'year without a summer' (Landsberg and Albert, 1974). The reconstruction shows anomalous low pressure over the whole map and is entirely consistent with the historical evidence of a cold, wet summer.

Some experimentation was tried using more complex models which incorporated lagged ring-width variables. These produced notable improvements in calibration explained variance but there was no corresponding increase in the verification performance. Frequently, this was worse than for the simple model. There is a strong tendency for complex models to overfit the calibration data even when noise reduction procedures are used to reduce the numbers of candidate predictors. Adopting a strict approach to verification of empirically tuned regression models ensures that a realistic appraisal of model strength and weaknesses is achieved.

THE FUTURE

Given the complexity and disjunct nature of weather over relatively small geographical areas and throughout the year, together with the great complexity of climatic and non climatic factors which affect tree growth on short and long timescales, it is perhaps surprising that any climate information can be gleaned through dendroclimatic studies in the British Isles. Nonetheless, objective calibration and verification of reconstruction models demonstrates that real information can be extracted using the ring widths of even a single genus. The results of work to date are certainly encouraging, the more so as these results were achieved using an extremely limited number of chronologies.

Our understanding of the physiology of tree growth and particularly the processes governing the growth of mature trees on a day to day basis throughout the growing season, remains limited. Progress in such research must also aid in the interpretation of previously produced empirical reconstructions and help to direct future work aimed at achieving optimum reconstruction performance.

Meanwhile, the empirical approach has much to offer. Given that Europe

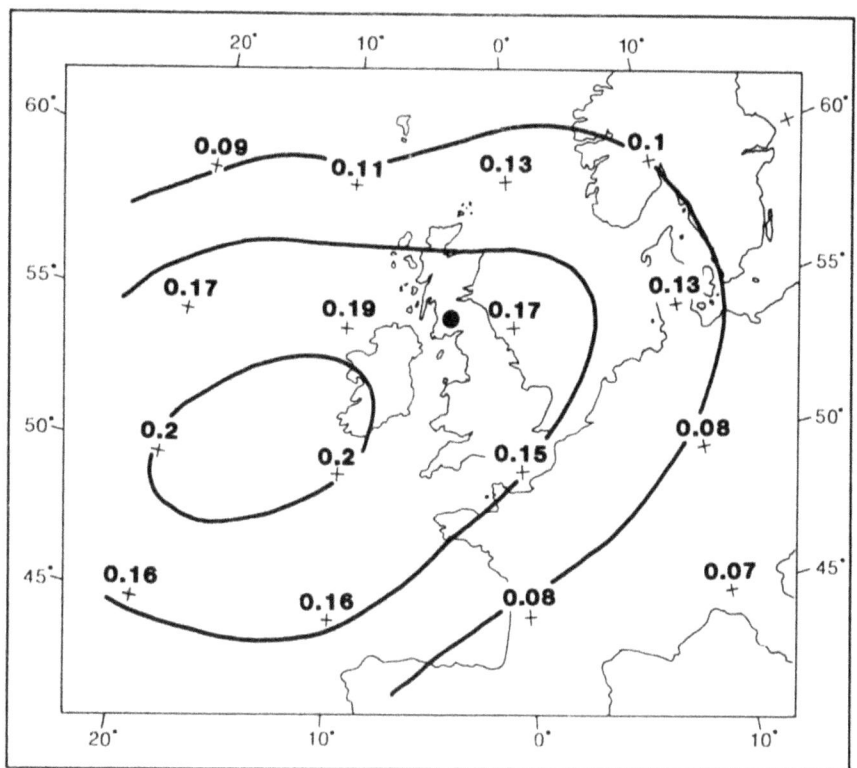

Figure 5. The map of the normalised regression weights for the Lockwood chronology (shown as a large dot) on each of the points in the 16 grid-point sea-level pressure reconstruction.

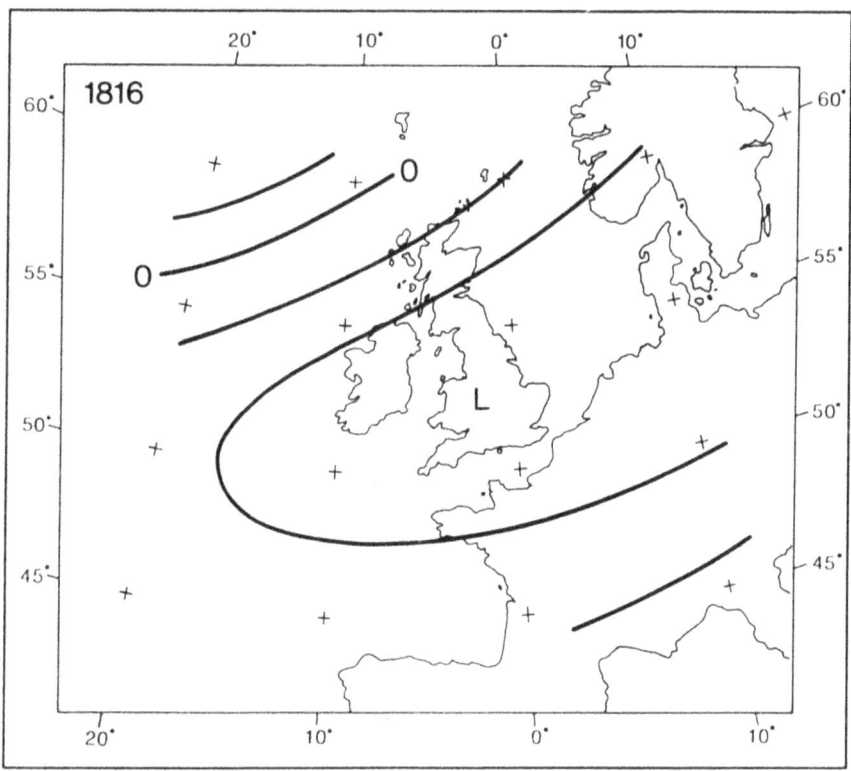

Figure 6. The ring-width based reconstruction of sea-level pressure for the 1816 summer season (May-July). The contours represent 0.5 mb anomalies from the 1873 to 1922 mean.

is particularly well blessed with some of the best and longest climate records to be found anywhere in the world, the immediate prospects for significantly extending this data base through dendroclimatology are limited. It is at the high latitude and high altitude sites where progress might be most readily forthcoming. These areas generally lack long climate records. They are also the areas where old living trees, perhaps 500 years old and more, can be found. The climate limitation of growth at such sites can be simple, direct and unambiguous and regression models should produce centuries of valuable data in these areas. More generally, the developments in the number, species representation and distribution of chronologies in Europe continues. An important development in recent years is the growth in the number of chronologies of densitometric data (Schweingruber et al., 1978; Hughes et al., 1984). These data are good proxies of summer temperatures and should make a valuable contribution to the dendroclimatic data base.

Of the various types of proxy climate data, tree-ring series may be limited in length but they are noteworthy for the high resolution information that they provide. Looking further to the future, there is the prospect of expanding the network of long composite chronologies of living, archaeological and sub-fossil material, such as that in Northern Ireland (e.g. Pilcher et al., 1984). If a sufficient number of these regional chronologies could be produced it might prove possible to extract annually resolved climate information on timescales of many hundreds or even thousands of years.

Meanwhile, European climate and dendrochronological data provide a testing ground for the development of reconstruction methods.

In other parts of the world there is rarely enough data to calibrate and verify empirical dendroclimatic models adequately. From one point of view the fact that many 'living' European chronologies extend back in time no further than the instrumental records may appear to be a disadvantage. In reality, the opposite is true. Statistical models are notoriously unreliable and strict validation of reconstruction, as noted by Fritts (1976) and Gordon (1982), is essential. Only in Europe are there enough instrumental data series to be able to do this satisfactorily. When this is done, faith can be placed in future reconstruction results.

REFERENCES

Bradley, R.S., Kelly, P.M., Jones, P.D., Diaz, H.F. and Goodess, C., 1985. A climatic data bank for the Northern Hemisphere land areas, 1851-1980. DoE Technical Report No. TR017, U.S. Dept. of Energy Carbon Dioxide Research Division, Washington, D.C., p.335.

Briffa, K.R., 1984. Tree-climate relationships and dendroclimatological reconstruction in the British Isles. Unpublished Ph.D. Dissertation, University of East Anglia, Norwich, England.

Briffa, K.R., Jones, P.D., Wigley, T.M.L., Pilcher, J.R. and Baillie, M.G.L., 1983. Climate reconstruction from tree rings: Part I, Basic methodology and preliminary results for England. Journal of Climatology, 3, pp. 233-242.

Briffa, K.R. Jones, P.D., Wigley, T.M.L., Pilcher, J.R. and Baillie, M.G.L., 1986. Climate reconstruction from tree rings: Part 2, Spatial reconstruction of summer mean sea-level pressure patterns over Great Britain. Journal of Climatology 6, 1-15.

Cook, E.R. and Jacoby, G.C., 1983. Potomac River Streamflow since 1730 as reconstructed by tree rings. Journal of Climate and Applied Meterology, 22, pp. 1659-1672.

Craddock, J.M. and Craddock, E., 1977. Rainfall at Oxford from 1767 to 1814, estimated from the records of Dr. Thomas Hornsby and others. The Meteorological Magazine 106, pp. 361-372.

Craddock, J.M. and Smith, C.G., 1978. An investigation into rainfall recordings at Oxford. The Meteorological Magazine 107, pp. 257-271.

Creber, G.T., 1977. Tree-rings: a natural data-storage system. Biological Reviews 52, pp. 349-383.

Fritts, H.C., 1976. Tree Rings and Climate. Academic Press, London, p. 567.

Fritts, H.C., Lofgren, G.R. and Gordon, G.A., 1979. Variations in climate since 1602 as reconstructed from tree rings. Quaternary Research 12, pp. 18-46.

Gordon, G.A., 1982. Verification of dendroclimatic reconstructions. (In) Climate from Tree Rings (Eds. M.K. Hughes, P.M. Kelly, J.R. Pilcher and V.C. LaMarche, Jr.). Cambridge University Press, Cambridge, pp. 58-61.

Gray, J., 1981. The use of stable-isotope data in climate reconstruction. (In) Climate and History: Studies in Past Climates and their Impact on Man. (Eds. T.M.L. Wigley, M.J. Ingram and G. Farmer). Cambridge University Press, Cambridge, pp. 53-81.

Holmes, R.L., Stockton, C.W. and LaMarche, V.C., Jr., 1979. Extensions of riverflow records in Argentina from long tree-ring chronologies. Water Resources Bulletin 15, pp. 1081-1085.

Hughes, M.K., Schweingruber, F.H., Cartwright, D. and Kelly, P.M., 1984: July-August temperature at Edinburgh between 1721 and 1975 from tree-ring density and width data. Nature 308, pp. 341-344.

Jacoby, G.C., (ed.), 1980. Proceedings of the International Meeting on Stable Isotopes in Tree-Ring Reseach and Assessment Program, Publication No. 12. U.S. Department of Energy, CONF-790518. UC-11, Department of Energy, Washington D.C.

Jarvis, M.G., 1973. Soils of the Wantage and Abingdon district. Mem. Soil Survey of Great Britain.

Jones, P.G., Wigley, T.M.L. and Briffa, K.R., 1983. Reconstructing surface pressure patterns using principal components regression on temperature and precipitation dates. Proceedings on the second International Meeting on Statistical Climatology, Instituto Nacional de Meteorologie e Geofisica, Lisbon, 4.2.1-4.2.8.

Jones, P.D., Briffa, K.R. and Pilcher, J.R., 1984. Riverflow reconstruction from tree rings in southern Britain. Journal of Climatology 4, pp. 461-472.

Kay, F.F., 1934. A soil survey of the eastern portion of the Vale of the White Horse. Bull. Fac. Agric. Hort. University of Reading No. 48.

Kozlowski, T.T., (Ed.) 1962: Tree-Growth. Ronald Press, New York, p. 442.

Kramer, P.J. and Kozlowski, T.T., 1962: Physiology of Trees. McGraw-Hill, New York.

LaMarche, V.C., Jr., 1978. Tree-ring evidence of past climatic variability. Nature 276, pp. 334-338.

Landsberg, H.E. and Albert, J.M., 1974. The summer of 1816 and volcanism. Weatherwise 27, pp. 63-66.

Long, A., 1982. Stable isotopes in tree rings. (In) Climate from Tree-Rings (Eds. M.K. Hughes, P.M. Kelly, J.R. Pilcher and V.C. LaMarche, Jr). Cambridge University Press, Cambridge, pp. 12-18.

MacCracken, M. C. and Luther, F.M, (Eds.) 1985: Detecting the economic effects of increasing carbon dioxide. U.S. Department of Energy CO_2 Research Program No. DOE/ER-0235. p. 198.

Manley, G., 1974: Central England temperatures: monthly means 1659 to 1973. Quarterly Journal of the Royal Meteorological Society 100, pp. 389-405.

Nicholas, F.J. and Glasspoole, J., 1931. General monthly rainfall over England and Wales, 1772-1931. British Rainfall (1931). pp. 299-306.

Osmond, D.A., Swarbrick, T., Thompson, C.R. and Wallace, T., 1949: A survey of the soils and fruits in the Vale of Evesham. Bull. Minist. Agric. Fis. Fd., London No. 116.

Pilcher, J.R., Baillie, M.G.L., Schmidt, B. and Becker, B., 1984. A 7,272-year tree-ring chronology for western Europe. Nature 313, pp. 150-152.

Robinson, K.L., 1948: The Soils of Dorset. (In) A geographical handbook of the Dorset flora. Dorset Nat. Hist. and Archaeol. Soc., County Museum, Dorchester.

Schweingruber, F.J., Fritts, H.C., Bräker, O.U., Drew, L.G. and Schär, E., 1978. The X-ray techniques applied to dendroclimatology. Tree-ring Bulletin 38, pp. 61-91.

Schweingruber, F.H., Bräker, O.U. and Schär, E., 1979. Dendroclimatic studies on conifers from central Europe and Great Britain. Boreas 8, pp. 457-452.

Smith, L.P. and Stockton, C.W., 1981. Reconstructed streamflow for the Salt and Verde Rivers from tree-ring data. Water Resources Bulletin 17, pp. 939-947.

Stockton, C.W., 1975. Long-term streamflow records reconstructed from tree-rings, Papers of the Laboratory of Tree-Ring Research No. 5. The University of Arizona Press, Tucson, U.S.A.

Stommel, H. and Stommel, E., 1979. The year without a summer. Scientific American 240, pp. 134-140.

Tabony, R.C., 1977. The variability of long-duration rainfall over Great Britain. Meteorological Office Scientific Paper No. 37, HMSO, London, p. 40.

Trenberth, K.A. and Paolino, D.A., 1980. The Northern Hemisphere sea-level pressure data set: trends errors and discontinuities. Monthly Weather Review 108, pp. 855-872.

Wigley, T.M.L., 1982. Oxygen-18, Carbon-13 and Carbon-14 in tree rings. (In) Climate from Tree Rings (Eds. M.K. Hughes, P.M. Kelly, J.R. Pilcher and V.C. LaMarche, Jr.). Cambridge University Press, Cambridge, pp. 18-21.

Wigley, T.M.L. and Tu Qipu, 1983: Crop-Climate Modelling Using Spatial Patterns of Yield and Climate. Part 1: Background and an example from Australia. Journal of Climate and Applied Meteorology 22, pp. 1831-1841.

Wigley, T.M.L., Lough, J.M. and Jones, P.D., 1984a: Spatial patterns of precipitation in England and Wales and a revised England and Wales precipitation series. Journal of Climatology 4, pp. 1-25.

Wigley, T.M.L., Briffa, K.R. and Jones, P.D., 1984b: On the average value of correlated time series with applications in dendroclimatology and hydrometeorology. Journal of Climate and Applied Meteorology 23, pp. 201-213.

Williams, J. and van Loon, H., 1976. An examination of the Northern Hemisphere sea-level pressure data set. Monthly Weather Review 104, pp. 1354-1361.

Wright, C.E., 1978. Synthesis of river flows from weather data. Technical Note No. 26, Central Water Planning Unit, Reading, U.K.

Zimmermann, M.H., (Ed.), 1964. The Formation of Wood in Forest Trees. Academic Press, New York.

Dendroclimatology of Pinus sylvestries L. in the Scottish Highlands

M.K. Hughes

Laboratory of Tree-Ring Research
University of Arizona,
Tucson, Arizona 85721
U.S.A.

ABSTRACT

The dendrochronological potential of Pinus sylvestris from the Scottish Highlands is explored for annual ring width and maximum latewood density. Between-site crossdating is good for ring width series from some sites and excellent for maximum latewood density in the majority of cases. The strong crossdating of maximum latewood density across the Highland region appears to be associated with a clear correlation between this variable and summer temperatures. An application of this finding to the reconstruction of past climate is discussed.

INTRODUCTION

Tree ring research in the British Isles differs from that elsewhere in its overwhelming concern with one tree genus, the oak. This has arisen from the preponderance of oak timber in artefacts and the relative lack of other native trees with suitable properties for dendrochronology, particularly longevity. Whilst this simplifies the conduct of dendrochronology, it also limits its interest and applicability, especially in the study of past environments and palaeoecology.

The other major candidate for tree ring research is the Scots pine Pinus sylvestris L., a species of importance over much of Europe at the present and of considerable significance in the vegetational history of the British Isles. A body of published work exists on the dendrochronology of Scots pine in Europe, for example from Scandinavia (Siren, 1961), France (Huber, 1976) and the Low Countries (Munaut, 1966). The work of Pilcher (1973) and McNally and Doyle (1984) on sub-fossil pines in Ireland are the only known examples of tree ring research on Scots pine in the British Isles using rigorous cross-dating on ring widths alone. Partial rings are more frequently encountered and the high correlation between the width of one year's ring and the next in many of the conditions found here can make sound cross-dating difficult.

In addition to its conservatism with regard to tree species, dendrochronology in the British Isles has been very largely limited to the use of radial increment (ring width) as the descriptor of the wood formed in a particular year. The best developed additional tree ring measurements are those derived from X-ray microdensitometry. This technique was first developed by Polge (1963), and further elaborated at the Swiss Forest Research Institute, Birmensdorf (Lenz et al., 1976). Carefully prepared wood samples are exposed to soft X-rays in such a manner as to produce high resolution radiographs on photographic film. Given the correct preparatory

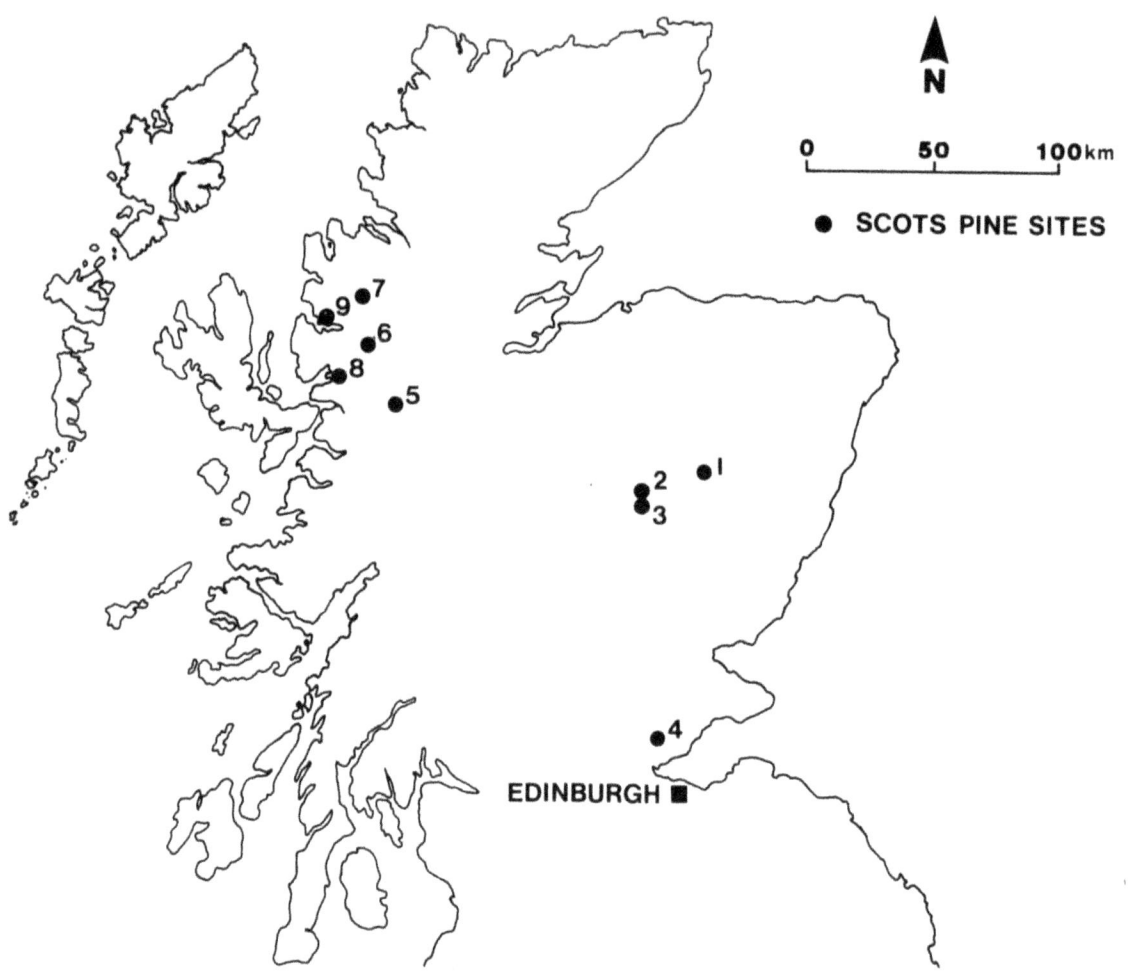

Figure 1. Map of distribution of sample sites. Numbers refer to Table 1.

TABLE 1: SITE DETAILS

Site	Latitude deg.min.N	Longitude deg.min.W	Altitude metres a.s.l.	Aspect	Slope deg.
1. Ballochbuie	56 57	03 19	381	NE	30–40
2. Mar	57 01	03 34	457	W	20–30
3. Inverey	57 00	03 35	500	W	15
4. Dimmie	56 07	03 20	200	–	–
5. Glen Affric	57 17	05 00	300	SE	10
6. Coulin	57 32	05 21	250	N	45
7. Loch Maree	57 31	05 21	100	NE	20–45
8. Plockton	57 20	05 38	100	W	5–10
9. Shieldaig	57 30	05 37	12	W	50

techniques and conditions of exposure and development, the optical density of the radiograph has a simple curvilinear relationship to the physical density of the wood. Calibration against standards of appropriate materials allows measurements of radiograph optical density made using an optical microdensitometer to be converted to wood specific gravity. This can be done with sufficient resolution to discern the details of within-year density variation. These methods are described in detail by Lenz et al. (1976) and Hughes and Sardinha (1975). Limited attempts have been made to use wood density as a dendrochronological parameter in work with British oak (Fletcher and Hughes, 1970; Milsom, 1979; Milsom and Hughes, 1978; Pilcher and Hughes, 1982), but the greatest promise of wood microdensitometry is with conifers. This has been well established by Polge (1978), Parker and Henoch (1971), Schweingruber et al (1978) and Conkey (1979). In particular, conifers of several genera have revealed strong common patterns in intra-annual density parameters at within-site and regional level, in particular the maximum density found in latewood (MXD). That is, MXD often shows strong cross-dating over great distances in conifers at high altitudes and latitudes (Schweingruber et al, 1979) and on occasion in less extreme conditions (Huber, 1976). Further, it has been demonstrated under a wide range of high altitude and high latitude conditions that intra-annual wood density measurements, especially MXD, are excellent recorders of past climate, especially summer temperature (Parker and Henoch, 1971; Schweingruber et al, 1978; Conkey, 1982).

It was in the context of these developments in technique that work was commenced on Pinus sylvestris in the Scottish Highlands. It seemed entirely reasonable to expect that this species would yield material suitable for the application of wood microdensitometry to tree ring studies, especially for the reconstruction of past climate. The aim was to explore the dendrochronology and dendroclimatology of Pinus sylvestris in the Scottish Highlands, using intra-annual density variables as well as ring width. The results of an extremely successful attempt at the reconstruction of past climate using the tree ring materials to be described here have been published (Hughes et al., 1984). There is a close, simple and temporally stable statistical relationship between tree ring variables (primarily MXD) in P. sylvestris in the Scottish Highlands and July/August mean temperatures. This paper gives an account of the materials used, their dendrochronological quality, inter-site cross-dating in ring width and MXD, site chronology climate responses and the relationships between the various chronologies derived from these materials, site conditions and regional climate.

MATERIALS

Table 1 gives the location and related site details for nine sites from which increment cores were taken from samples of the P. sylvestris population. They span the Scottish Highlands from strongly oceanic conditions in the west to more nearly continental in the east (Figure 1). Some, for example Ballochbuie and Inverey, are close to the potential tree line whilst others such as Loch Maree are well below this. All are on relatively impoverished soils, but with varied aspect and slope. Where trees of more than 180 years age are found stand structure is in general open with a preponderance of older trees, except at Glen Affric. All sites but Dimmie and Plockton are in native pinewoods documented by Stevens and Carlisle (1959) or Bunce (1977). Increment cores were taken, conditioned, prepared and analysed using the methods described by Lenz et al. (1976). X-

TABLE 2: CHRONOLOGY STATISTICS

SITE	1. Balloch	2. Mar	3. Inverey	4. Dimmie	5. Affric	6. Coulin	7. Loch Maree
Chronology period	1712-1975	1773-1977	1706-1975	1828-1975	1735-1975	1671-1977	1765-1977
Number of trees*	11	13	12	11	12	10	8
RING WIDTH							
Serial correlation	0.58	0.45	0.73	0.64	0.78	0.69	0.54
Mean sensitivity	0.13	0.13	0.12	0.16	0.13	0.13	0.14
Mean value (1/100mm)	117	119	109	168	110	100	136
Standard deviation	0.17	0.15	0.21	0.20	0.27	0.21	0.18
%Y	39	19	31	32	13	24	35
MXD							
Serial correlation	0.05	-0.04	0.43	0.45	0.36	0.22	0.24
Mean sensitivity	0.08	0.07	0.06	0.06	0.03	0.06	0.05
Mean value (10^{-2} mg/ml)	79	79	79	79	94	84	80
Standard deviation	0.07	0.06	0.07	0.08	0.04	0.06	0.05
%Y	54	51	38	21	30	36	34

%Y is the proportion of sample variance retained by the final chronology as determined from an analysis of variance over the 60 year period 1910-1969.

A full explanation of the calculation and intepretation of chronology statistics is given by Fritts (1976).

* two cores were taken from each tree.

ray densitometry and associated preparation was done at the Swiss Forest Research Institute, Birmensdorf under the supervision of F.H. Schweingruber. Cross-dating of all cores was checked there graphically using graphs of ring width, early and latewood widths, minimum and maximum density (MXD). These raw data were then sent to Liverpool where a further check on cross-dating was combined with the use of the program CROS which uses sliding comparisons of Student's 't' in checking positions of match (Baillie and Pilcher, 1973). It was generally the case that cross-dating was less ambiguous for MXD than ring width. For example, at the Inverey site the mean level of 't' between pairs of series within trees was 11.52 for MXD and 8.55 for ring width. The corresponding values for between-tree comparisons were 11.71 for MXD and 7.43 for ring width. All these are for overlaps of at least 200 years. The series of measurements of ring width and MXD for the cross-dated cores were then standardised to allow for the effects of age on tree growth. This was done by fitting an orthogonal polynomial to each series and taking the quotient of the actual and calculated values as the index. These individual core series were then averaged for each site to produce site chronologies for ring width and MXD (Fritts, 1976). Hence the analyses reported here concern a maximum of eighteen site chronologies, nine each for ring width and MXD.

CHRONOLOGY STATISTICS

It is possible to describe a number of the properties of a site chronology in relatively simple statistical terms. This is of particular value in comparing series from differing regions or species as well as those derived from various tree ring measurements such as ring width and MXD. Table 2 gives such statistics for ring width and MXD chronologies for seven of the sites, all analyses of variance being for the period 1911-1970. Serial correlations are markedly higher in the ring width series than in MXD. Whilst the highest %Y (the proportion of variance attributable to the common between-year pattern) was for an MXD series and the lowest for a ring width series, there was an overlap between values of this parameter between the two tree ring variables. There is no consistent relationship between either serial correlation or %Y and the mean value of ring width or MXD. Whilst the highest values of %Y for MXD are found at high altitude sites in the East, there is no consistent relationship between chronology statistics and altitude or continentality.

The serial correlation values for ring width chronologies are similar to or higher than those for oak chronologies in the British Isles (Hughes et al., 1978; Pilcher and Baillie, 1980a, 1980b). %Y for Scots pine ring width tends to be lower than for oak. Whilst serial correlation for MXD is smaller than for ring width in Scots pine, it is not generally as small as in the alpine conifers reported by Schweingruber et al. (1978). The main distinguishing feature of the MXD chronology statistics is the very much lower serial correlation found in these series as compared to ring width from the same samples. This indicates a system with lower persistence, a positive attribute in a proxy record of past environment. This results in a more faithful recording of high frequency environmental variation by MXD and a more tractable time series for statistical analysis.

Inter-Site Cross-Dating

Pairwise comparisons were made between the nine site chronologies for

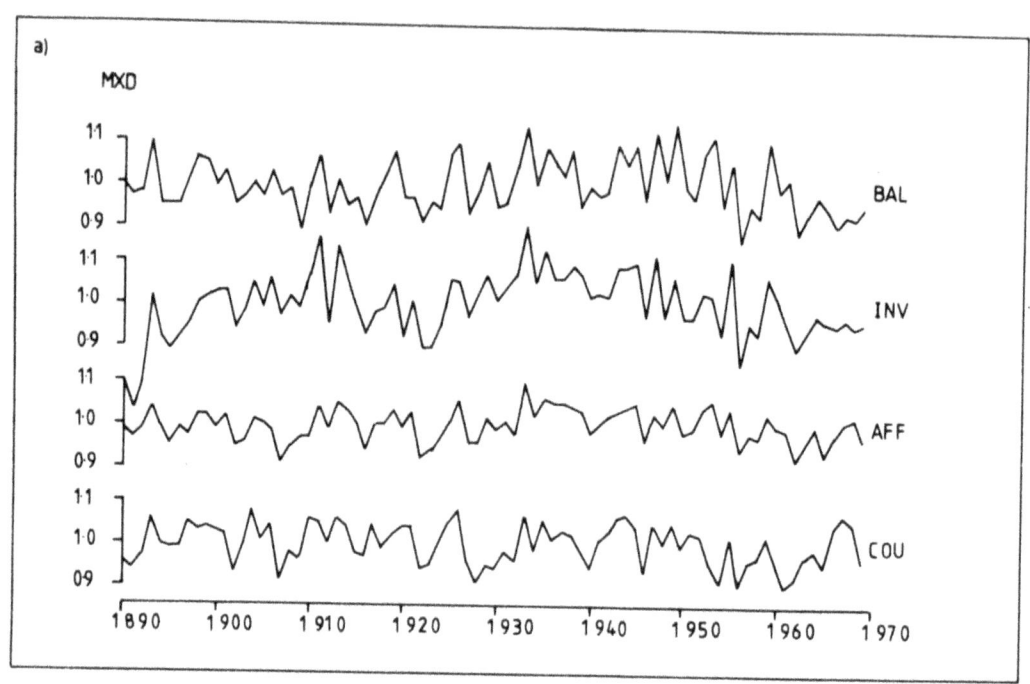

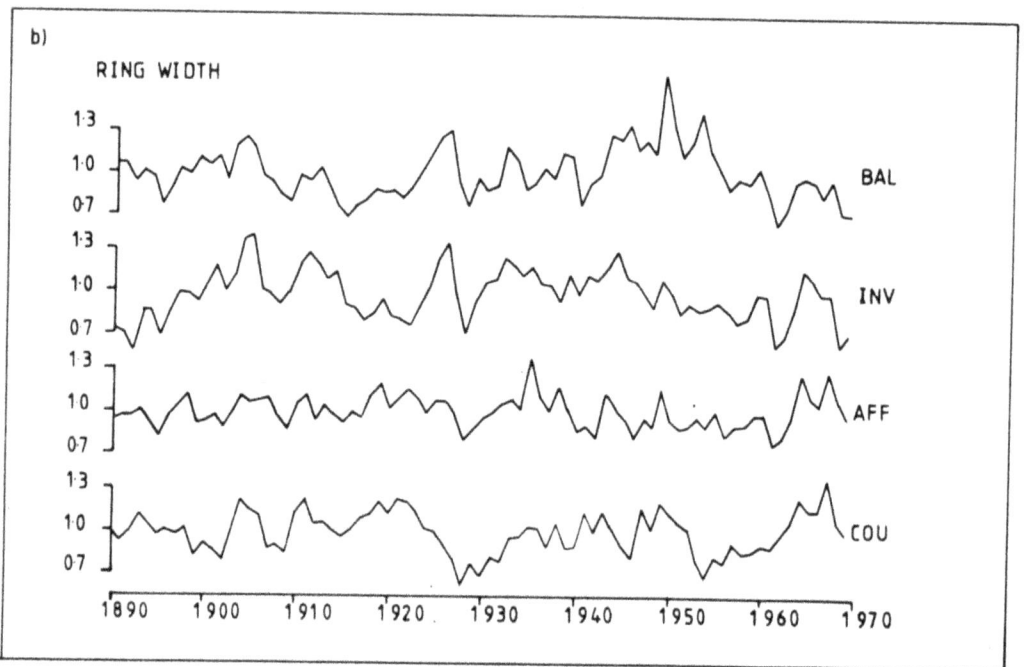

Figure 2. (a) Indexed maximum latewood density (MXD) chronologies from four sites. BAL = Ballochbuie; INV = Inverey; AFF = Glen Affric; COU = Coulin.
(b) Indexed ring width chronologies from the same four sites.

both ring width and MXD using program CROS (Baillie and Pilcher, 1973). Table 3 gives results as 't' values for the 97 year period 1879-1975. Values for MXD comparisons are always higher than for ring width, there being several cases of very good match in MXD but little or no evidence of a match between ring width chronologies. As figure 2 confirms, excellent matches exist between MXD series right across the Highlands. There is cross-dating in the ring width chronologies, comparable to many between oak chronologies over similar distances (Pilcher and Baillie, 1980a), but inferior to that found in MXD from the same pine. These results are evidence for a regional scale environmental factor or complex of factors affecting the processes determining Scots pine MXD. Trees growing in extremely disparate sites contain the same interannual pattern of variation in MXD, largely independent of site conditions.

Climatic Response

The strongest candidate for the role of external factor controlling MXD variation must be climate. It is difficult to see what other factors could act in such a consistent way over such large distances and long times. In order to test this, climate response functions have been calculated for the MXD and ring width chronologies using the method of Fritts et al. (1971). Monthly precipitation and mean temperature data from nearby meteorological stations are used as predictors of ring width index or MXD index in an orthogonalised regression analysis using several decades' tree ring data and meteorological records. Persistence in the tree ring series is modelled by forcing one or more lagged values of the predictand into the predictor set. The outcome of the analysis is expressed in two principal forms. Firstly, the proportion of ring width or MXD chronology variance accounted for can be expressed as the percentage of chronology variance accounted for by the regression model (Table 4). Secondly, the importance of each monthly value of temperature or precipitation as a predictor is given by a response function element. It is thus possible to compare the form of response functions for a variety of site chronologies. This admittedly crude statistical model of tree ring climate relationships is remarkably effective as a predictor of MXD site chronologies, especially at the higher altitude sites in the eastern Highlands (Table 4). This model of climate is a less effective predictor of ring width site chronologies, although its performance is better at most of the higher altitude, eastern sites than in the west. Prior growth is an important predictor in all the ring width chronologies except Glen Affric, and rarely so in the MXD chronologies. There is little consistency in the response functions for ring width except that all have a significantly negative element for the prior August's temperatures (Figure 3), in contrast to MXD (Figure 4) where all sites have significantly positive response function elements for current summer temperatures. The sites in the Cairngorm region also have significantly positive elements for spring temperatures.

These unambiguous correlations between MXD and summer temperature provide an explanation for the very strong regional pattern of variation in MXD, clearly over-riding site differences. Temperature varies in a spatially coherent manner especially when averaged over periods of several weeks, even in a mountainous region such as the Scottish Highlands. This is likely to be of particular significance to the control of latewood density in P. sylvestris, since this is probably determined in large part by cell wall growth over an extended period in the current summer. The production of earlywood plays an important part in the determination of ring

TABLE 3: CROSS-DATING OF SCOTS PINE - 't' VALUES

MXD

SITE	1. BAL	2. MAR	3. INV	4. DIM	5. AFF	6. COU	7. LMA	8. PLO	9. SHL
1. BAL	–	21.40	13.13	7.12	9.48	10.26	9.48	2.78	6.52
2. MAR	10.71	–	18.06	7.38	15.43	14.84	13.82	4.66	8.67
3. INV	7.38	8.86	–	7.61	11.34	10.25	7.80	4.81	6.22
4. DIM	5.48	4.12	0	–	7.07	5.46	5.42	4.78	5.35
5. AFF	5.76	8.53	6.75	4.00	–	10.23	8.03	5.54	6.95
6. COU	5.84	7.62	4.79	2.29	8.32	–	16.93	5.16	7.79
7. LMA	0	6.58	3.13	4.43	8.28	6.28	–	7.63	11.24
8. PLO	0.81	0.08	0	0	0	0.33	0.91	–	9.37
9. SHL	0	4.72	2.37	2.67	4.95	2.89	5.50	0	–

RING WIDTH

The upper triangle gives values for MXD, the lower for ring width. All comparisons are for the 97-year period 1879-1975. 't'-values were calculated using program 'CROS' (Baillie and Pilcher, 1973).

TABLE 4: % VARIANCE ACCOUNTED FOR BY RESPONSE FUNCTIONS

	RING WIDTH		MXD	
SITE	Climate	Prior growth	Climate	Prior growth
1. Ballochbuie	39.7	33.1	68.8	3.4
2. Mar	38.6	31.8	63.2	13.1
3. Inverey	11.6	58.2	58.8	18.3
4. Dimmie	32.3	48.4	59.8	14.3
5. Glen Affric	20.6	16.6	65.3	6.7
6. Coulin	22.1	57.5	49.6	4.2
7. Loch Maree	2.3	75.9	42.7	21.9
8. Plockton	28.3	31.6	37.6	0

All response functions calculated for the 60 years 1911-1970.

Braemar temperature and precipitation were used for sites 1 to 4, Achnashellach temperature and Portree precipitation for sites 5 to 8.

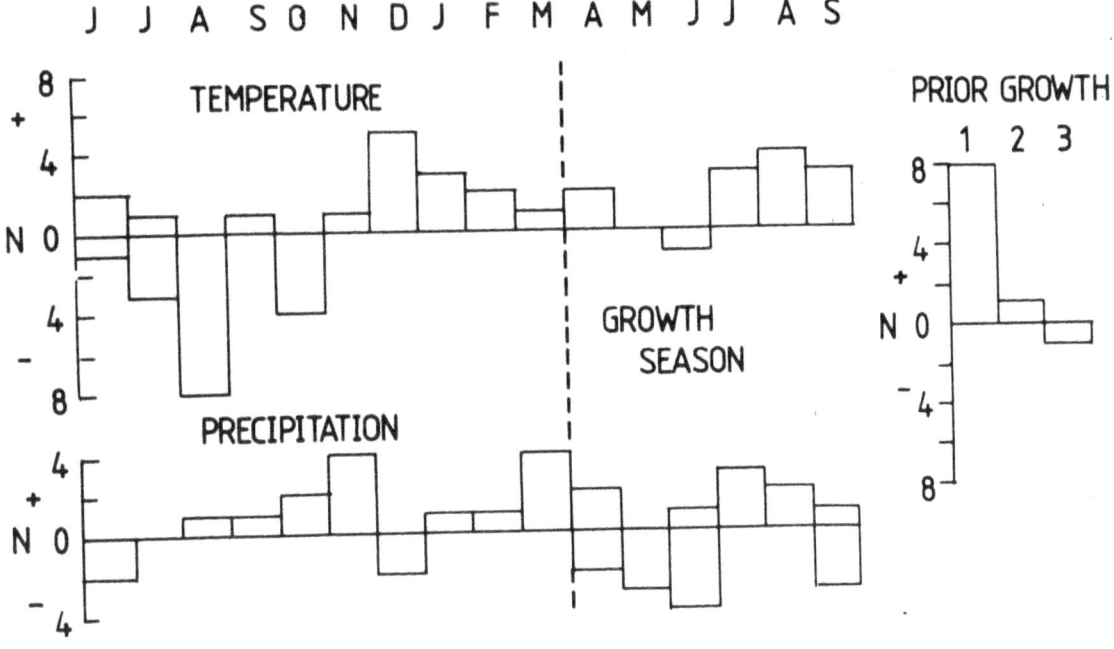

Figure 3. Summary response function (Pilcher and Gray, 1982) for ring width. N is the number of positive (+) or negative (−) response function elements for each month and variable from the 8 response functions listed in Table 4. Counts are also given for prior growth at lags 1 to 3.

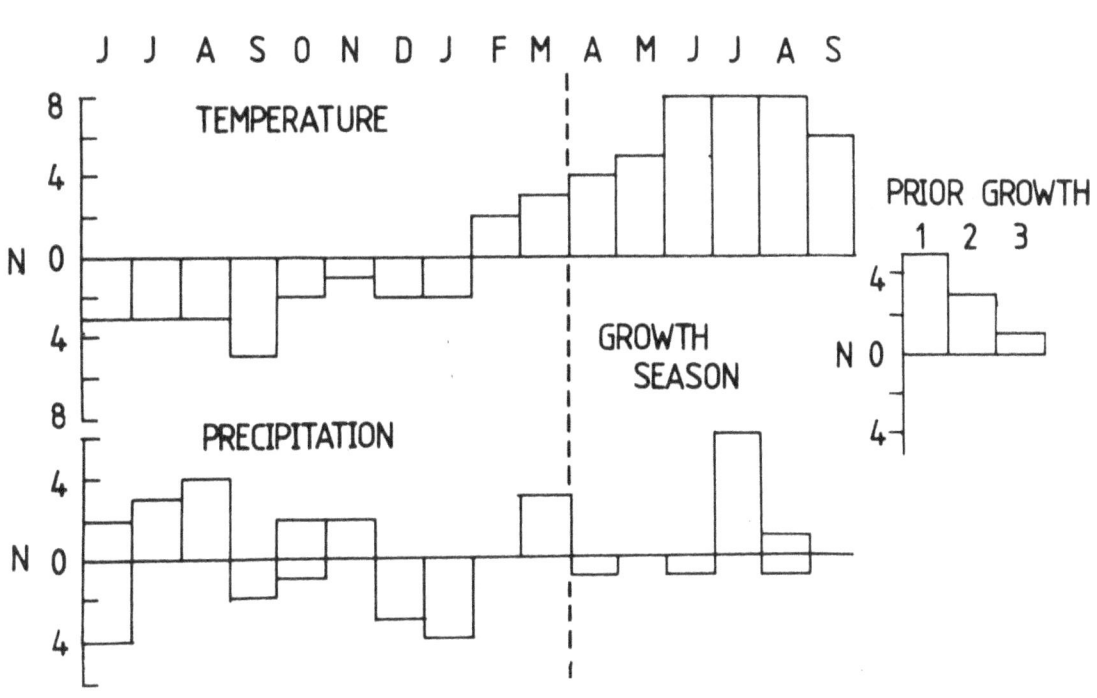

Figure 4. Summary response function for maximum latewood density (MXD).

width. In contrast to MXD this is influenced not only by current conditions but also by the storage of carbohydrates in parenchyma cells during the preceding late summer. Thus it is to be expected that earlywood width and hence ring width will display a higher serial correlation because of its relation to events in the previous year.

CLIMATE RECONSTRUCTION

Hughes et al. (1984) demonstrated that it was possible to reconstruct past temperature at Edinburgh using tree-ring density (MXD) and ring width data from five of the sites described here. The climate-growth relationship was formalised in an equation that allowed the reconstruction of a climate variable from the tree ring data. This procedure is known as calibration. Its efficacy was then tested (verified) by comparing the reconstructed climate data with actual climate data for a period of years not used in the calibration. In this case the predictors were the five pairs of MXD and ring width chronologies. The tree ring data for year i - 1, and i + 1 were presented to the regression as well as those for year i, the current year, so as to allow for autoregression in the tree ring data and for lag effects. The predictand was the mean July-August temperature at Edinburgh. This station was chosen because records are available continuously from 1764 to the present, providing an almost unequalled opportunity to test the techniques of dendroclimatology. Calibration and verification was carried out in two stages. In the first stage the 160-year period 1810-1969 was divided into two 80-year periods, A and B. Period A was first used for calibration, and period B for verification. Then their roles were reversed, so that the temporal stability of the regression model and its performance could be tested. The results were good on both counts, 60% of temperature variance being accounted for in the independent, verification periods and the same predictors making the major contribution with similar regression coefficients in both cases. These were current year MXD at Inverey and Ballochbuie and ring width in year i+1 at Ballochbuie. The second stage was to use the whole 160-year period for calibration, and the 45 years 1765 to 1809 for verification. The new model accounted for 48% of temperature variance in this verification period, even though the instrumental record for the early part of the period, especially 1777 to 1783, is open to some doubt (see Hughes et al., 1984). The structure of the model was similar to those derived for the 80-year periods, with the addition of MXD at Coulin as an important predictor.

The structure of the regression models reported suggests that much of the success of this attempt to reconstruct past July-August temperature at Edinburgh arises from the clear statistical relationship between MXD at all the sites reported and summer temperatures, as shown by response function analysis (Figures 3 and 4). This can be tested by repeating the calibration and verification procedures using either MXD or ring width data alone. Table 5 gives the results of this approach applied to MXD, along with those already published for MXD and ring width used together by Hughes et al (1984). Calibrations were also attempted for ring width alone against July-August temperature. In no case did variance accounted for by the regression model exceed 17% in the calibration period. Consequently no reconstruction or verification was attempted. The calibration and verification statistics for MXD alone in the two 80-year periods are generally poorer than for MXD and ring width used together (Table 5), although still creditable. The MXD models show the same temporal stability. The relative performances of the 160-year calibrations were

TABLE 5: CALIBRATION AND VERIFICATION STATISTICS FOR RECONSTRUCTIONS BASED ON MXD CHRONOLOGIES (MXD + RING WIDTH IN PARENTHESES) *

CALIBRATION PERIOD	1890-1969	1810-1890	1810-1969
Mean of actual values (deg. C)	14.46	14.39	14.42
Std. deviation of actual values (deg. C)	0.83	0.88	0.85
Multiple correlation coefficient (1)	0.70 (0.76)	0.58 (0.63)	0.64 (0.69)
Serial correlation of residuals	-0.004 (-0.10)	0.22 (0.18)	0.11 (0.05)
No. of predictors retained	4 (7)	2 (3)	8 (9)
VERIFICATION PERIOD	1810-1889	1890-1969	1765-1809
Mean of reconstructed values (deg. C)	14.55 (14.41)	14.39 (14.41)	14.89 (14.50)
Correlation coefficient (1)	0.64 (0.78)	0.59 (0.79)	0.52 (0.69)
Reduction of error (RE)	0.25 (0.37)	0.38 (0.42)	0.16 (0.01)
Sign test (agreements/total) (2)	60/79 (63/79)	57/79 (59/79)	30/44 (31/45)
Product means test ('t' value) (3)	5.11 (4.75)	4.36 (4.82)	1.78 (2.73)

* A clear explanation of the statistics used is given by Fritts (1976)

(1) - all significant at p 0.001

(2) - all significant at p 0.01

(3) - all significant at p 0.001 for the 80-yr periods, only (MXD + ring width) at p 0.01 for the 45-yr period.

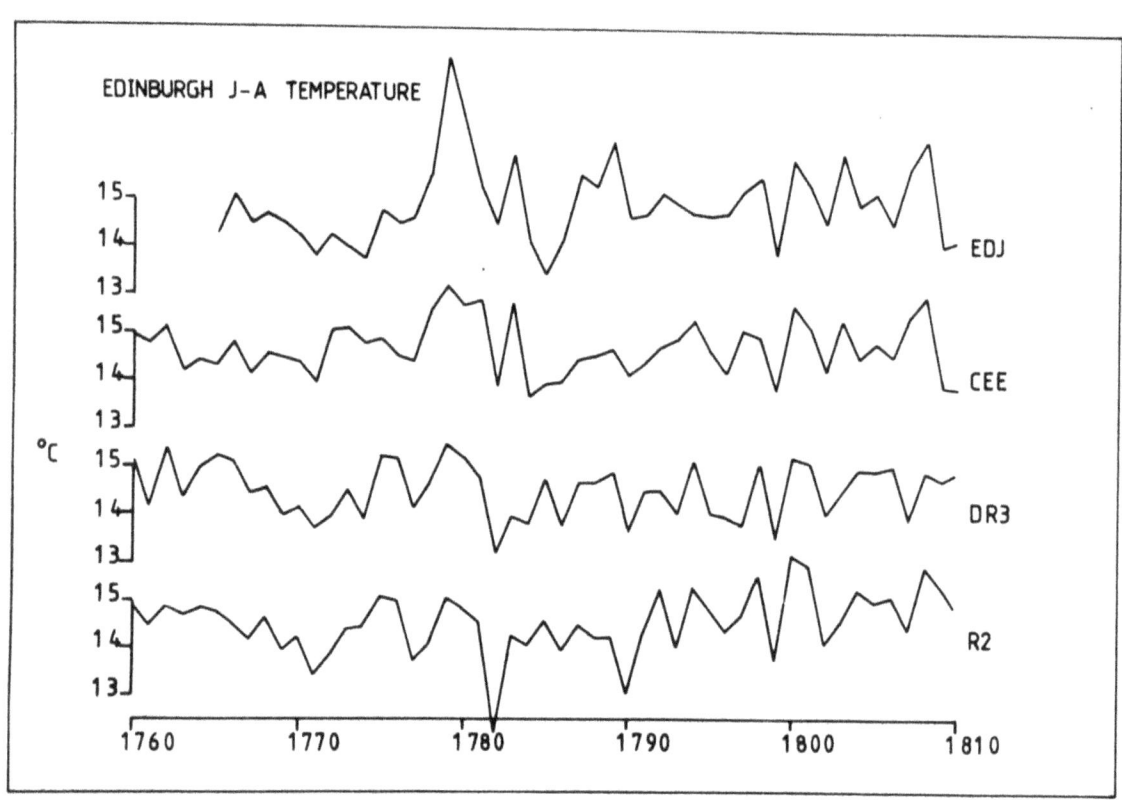

Figure 5. Edinburgh July-August mean temperature, 1760-1810.
EDJ = Edinburgh instrumental record
CEE = Manley's Central England record reduced to Edinburgh (See Hughes et al, 1984)
DR3 = as reconstructed using MXD only
R2 = as reconstructed using (MXD + ring width)
Both reconstructions are derived from calibrations determined for the period 1810-1969.

similar in all but one important respect. The MXD alone reconstruction had a reduction of error (RE) statistic of 0.16 compared with 0.01 for the MXD and ring width reconstruction, both calculated for the independent verification period 1765-1809. This is an important difference since any value of RE greater than zero indicates some skill in the model compared with the trivial persistence estimate. It appears that MXD alone more accurately reconstructs the mean temperature for 1765-1809. It may be that the single ring width chronology used in the MXD and ring width reconstruction produces a distortion in this period during which it is derived from relatively few trees. The two reconstructions are shown with the instrumental data in Figure 5.

CONCLUSIONS

Pinus sylvestris material from a range of sites in the Scottish Highlands has been shown to have considerable potential for dendrochronological use. Between-site cross dating is good for ring width in several cases and excellent for MXD in the majority of cases. The most notable difference between the MXD site chronologies and British Isles ring width chronologies, whether of Scots pine or oak, is the former's relatively small serial correlation. It was not possible to show a consistent relationship between the statistical properties of the chronologies and site conditions, although the greatest common variance (%Y) was found at high elevation sites in the Cairngorm area. The strong cross dating of MXD across the Highland region appears to be associated with a clear correlation between MXD and summer temperatures at all the sites reported here, as revealed by response function analysis. Monthly temperature variations tend to behave in a spatially coherent manner. No extensive clear and consistent climatic association was found for the ring width chronologies.

The potential of this species as a source of proxy climate information has already been demonstrated. MXD has been shown to contribute the major part of the summer temperature related information, although the inclusion of ring width series improves the performance of reconstructions in general terms. This is consistent with the findings of, for example, Schweingruber et al (1978) in the case of subalpine conifers in Switzerland.

The use of X-ray microdensitometry to extend the information extracted from the annual rings of Scots pine greatly increases the potential of this material, at least under conditions such as those found in much of the Scottish Highlands. The most immediate application is in the dendroclimatological reconstruction of past climate over a longer period using timber from peat and from artefacts. The existing chronologies will contribute to networks of tree ring chronologies currently being compiled as part of a European project in dendroclimatology and of a Northern Hemisphere project. The results reported here also indicate the dating potential of this material, although it should be stressed that a relatively regular wood structure is needed for X-ray microdensitometry and that the method is not suited to the analysis of extremely small rings.

Acknowledgements

The densitometric analyses were carried out under the supervision of Dr. F.H. Schweingruber at the Swiss Federal Forest Research Institute, Birmensdorf. I am indebted to him and his colleagues. Field collection was conducted by Dr. Schweingruber and David Cartwright, who also carried out certain statistical analyses. Liverpool Polytechnic supported this project. Permission to take core samples was most generously given by the Factor of the Balmoral Estate, the Forestry Commission, the Nature Conservancy Council and a number of landowners.

REFERENCES

Baillie, M.G.L. & Pilcher, J.R. 1973. A simple cross-dating program for tree-ring research. Tree-ring Bulletin, 33, pp. 7-14.

Bunce, R.G.H. 1977. The range of variation within the pinewoods. In: Native Pinewoods of Scotland, eds. R.G.H. Bunce and J.N.R. Jeffers. I.T.E. Cambridge.

Conkey, L.E. 1979. Response of tree-ring density to climate in Maine, U.S.A. Tree-ring Bulletin 39, pp. 29-38.

Conkey, L.E. 1982. Temperature reconstructions in the northeastern United States. In: Climate from tree rings, eds. M.K. Hughes, P.M. Kelly, J.R. Pilcher & V.C. LaMarche, C.U.P.: Cambridge.

Fletcher, J.M. & Hughes, J.E. 1970. Uses of X-rays for density determinations and dendrochronology. In: Tree ring analysis with special reference to N.W. America. Univ. Br. Colum. Publs. Fac. For. Bull, 7, pp. 41-54.

Fritts, H.C. 1976. Tree rings and climate, Academic Press: London.

Fritts, H.C., Blasing, T.J., Hayden, B.P. & Kutzbach, J.E. 1971. Multivariate techniques for specifying tree growth and climatic relationships and for reconstructing anomalies in paleoclimate. J. appl. Meteorol., 10, pp. 845-864.

Huber, F. 1976. Problemes d'interdatation chez le pin sylvestre et l'influence du climat sur la structure de ses accroissements annuels. Ann. Sci. For., 33, pp. 61-86.

Hughes, J.R. & Sardinha, R.M. de A. 1975. The application of optical densitometry in the study of wood structures and properties. J. Micros., 104, pp. 91-103.

Hughes, M.K., Gray, B., Pilcher, J.R., Baillie, M.G.L. & Leggett, P. 1978. Climatic signals in British Isles tree ring chronologies. Nature, Lond. 272, pp. 605-606.

Hughes, M.K., Schweingruber, F.H., Cartwright, D. & Kelly, P.M. 1984. July-August temperature at Edinburgh between 1721 and 1975 from tree-ring density and width data. Nature, Lond. 308, pp. 341-344.

Lenz, O., Schaer, E. & Schweingruber, F.H., 1976. Methodische probleme bei der radiographisch - densitometrischen Bestimmung der Dichte und der Jahrringbreiten von Holz. Holzforschung, 30, pp. 114-123.

McNally, A. and Doyle, G.J. 1984. A study of subfossil pine layers in a raised bog complex in the Irish Midlands. I. Palaeowoodland extent and dynamics. Proceedings of the Royal Irish Academy, 84, pp. 57-70.

Milsom, S.J., 1979. Within- and between - tree variation in certain properties of annual rings of sessile oak, Quercus petraea (Mattuschka) Liebl. as a source of dendrochronological information. Ph.D. thesis, C.N.A.A.

Milsom, S.J. and Hughes, M.K. 1978. X-ray densitometry as a dendrochronological technique. In: Dendrochronology in Europe, ed. J. Fletcher. B.A.R. International Series, 51, pp. 317-324.

Munaut, A.V., 1966. Recherches dendrochronologiques sur Pinus sylvestris I. Etude de 45 pins sylvestres recents originaires de Belgique. Agricultura, 2nd series, 14, pp. 193-292.

Parker, M.L. & Henoch, W.E., 1971. The use of Engelmann spruce latewood density for dendrochronological purposes. Can. J. For. Res. 1, pp. 90-98.

Pilcher, J.R., 1973. Tree-ring research in Ireland. Tree-ring Bulletin, 33, pp. 1-5.

Pilcher, J.R., & Baillie, M.G.L. 1980a. Eight modern oak chronologies from England and Scotland. Tree-ring Bulletin, 40, pp. 45-58.

Pilcher, J.R., & Baillie, M.G.L. 1980b. Six modern oak chronologies from Ireland. Tree-ring Bulletin, 40, pp. 23-34.

Pilcher, J.R. & Gray, B. 1982. The relationships between oak tree growth and climate in Britain. J. Ecol., 70, pp. 297-304.

Pilcher, J.R. & Hughes, M.K. 1982. The potential of dendrochronology for the study of climate change. In: Climatic change in late prehistory, ed. A.F. Harding. Edinburgh University Press.

Polge, H. 1963. Une nouvelle methode de determination de la texture du bois par l'exploration densitometrique de cliches radiographiques. Ann. Ec. Natl. For. st. Rech. Exper., 20, pp. 531-581.

Polge, H. 1978. The contribution of wood density to dendrochronology and dendroclimatology. In: Dendrochronology in Europe, ed. J. Fletcher. B.A.R. International Series, 51, pp. 77-87.

Schweingruber, F.H., Fritts, H.C., Bräker, O.U., Drew, L.G. & Schär, E. 1978. The X-ray technique as applied to dendroclimatology. Tree-ring Bulletin, 38, pp. 61-91.

Schweingruber, F.H., Bräker, O.U. & Schär, E. 1979. Dendroclimatic studies on conifers from central Europe and Great Britain. Boreas, 8, pp. 427-452.

Siren, G. 1961. Skogsgranstalle som indikator for klimafluktuationaerna i norra Fennoskandien under historisk tid. Comm. Inst. For. Fenn., 54.

Steven, H.M. & Carlisle, A. 1959. The Native Pinewoods of Scotland. Oliver & Boyd: Edinburgh.

Dendroglaciological Investigations in Norway

J.L. Innes

Forestry Commission
Forest Research Station
Alice Holt Lodge
Wrecclesham
Farnham
Surrey GU10 4LH

ABSTRACT

Dendroglaciology provides a means of reconstructing the history of past glacier fluctuations over periods of several hundred years in various parts of the world. The technique has not yet been fully developed and a considerable amount of work is required on its methodology. Norway represents a suitable environment for examining the technique in view of the detailed glaciological records and the past history of tree-ring work. A sampling network of over 40 sites has been established using Pinus sylvestris L. The potential of these for dendroglaciological studies is currently being examined in detail.

INTRODUCTION

Dendroglaciology is the use of tree-rings to reconstruct past glacier fluctuations. Successful applications of the technique have been carried out in the European Alps (e.g. LaMarche and Fritts 1971, Heikkinen 1980, Bircher 1982, Renner 1982), the USSR (Adamenko 1963, Lovelius 1972), North America (Bray and Struik 1963, Heikkinen 1984) and Norway (Matthews 1976, 1977a). The technique should be distinguished at the outset from the use of fossil tree trunks incorporated into moraines to date past glacial advances (e.g. Schneebeli 1976, Röthlisberger 1976) and from the use of living trees growing on moraines or on supraglacial debris (e.g. Sigafoos and Hendricks 1961, Giardino et al. 1984). The technique is also separate from attempts to relate evidence of past tree-ring changes to glacier fluctuations (e.g. Karlén 1976, Denton and Karlén 1977).

The basic premise underlying dendroglaciology is that the climatic parameters that control tree growth in an area also control the size of any glaciers that might be present. At high altitudes or high latitudes, the primary control on tree growth is the summer temperature (see below). This is also the main control on the mass balance of many continental glaciers, as it determines the extent of summer ablation (e.g. Paterson 1981, Haakenson and Roland 1984). Consequently, tree-ring studies may provide a means of reconstructing past mass balances and, therefore, by extension, glacier fluctuations. To do this, it is necessary to have good mass balance data, a number of reliable tree ring chronologies and, preferably, a good meteorological data base. Some of the most reliable mass balance data in the world have been collected by the research programmes of the Norwegian Polar Institute and the Norwegian Water Resources and Electricity Board

(NVE). Norway has also a tradition of dendrochronological and dendroecological research with the result that several long tree-ring chronologies exist. In addition, there is an extensive network of meteorological stations, several of which are situated at high altitudes, relatively close to glaciated areas. This makes Norway an ideal location for the investigation of dendroglaciological techniques.

This paper presents the background to, and rationale of, a dendroglaciological project currently underway in Norway and outlines previous work in the region relevant to it.

DENDROGLACIOLOGICAL STUDIES OF MATTHEWS

Matthews (1976, 1977a) has attempted to use dendroglaciology to determine the past history of Storbreen. Both papers are based on the data presented by Slåstad (1957) and Liestøl (1967). The chronology used was that derived for upper Gudbrandsdalen, which extends up to 1950. The Storbreen mass balance measurements were started in 1949 which leaves 1 year for the calibration with mass balance. This is obviously unsatisfactory. One way to surmount this problem is to use the reconstructed mass balance data presented by Liestøl (1976), which is based on the meteorological data from Fanaråken and Bergen. However, this introduces a potential source of error and is inadvisable, especially when using the data for calibration. A further problem is that Gudbrandsdalen and Storbreen are some distance apart (the Nord Fron site is 80 km away) which may influence any climatic relationship between the two. Despite these problems, Matthews managed to show that periods of reduced tree growth could be correlated with periods of glacial advance (as indicated by the presence of moraines), suggesting that the technique may be used successfully in some Norwegian situations.

RECENT GLACIAL FLUCTUATIONS IN NORWAY

There are two approaches to establishing the nature of recent glacier fluctuations. Firstly, changes in length can be directly measured or inferred from the positions of terminal and recessional moraines. Secondly, mass balance measurements can be made and the data used to calculate volumetric changes. Both techniques have been used successfully in Norway although by far the longest records exist for glacier lengths. Numerous measurements were made by early geologists and explorer-scientists (e.g. Kaiser Wilhelm II, Marstrander, the Prince of Monaco, Rekstad, Rabot, Richter, de Seue, and others). These form an extremely useful basis for more recent studies. Much of the earlier work has been summarised by Faegri (1950). Since then, the annual reports of the NVE and the Norwegian Polar Institute have documented the changes in the lengths of selected glaciers.

Data on frontal positions can also be obtained indirectly by dating recessional moraines. In Norway, the majority of dating has been by lichenometry, although there are a number of problems associated with the technique (Innes 1984). The best data are available for Storbreen in the Jotunheimen massif, where Matthews (1974, 1975, 1977b) has conducted detailed lichenometric investigations. Other studies in Norway, such as those of Andersen and Sollid (1971) and Karlén (1979) are much less reliable due to difficulties over the establishment and application of the lichenometric dating curve. The potential errors involved in the dating of

Table 1. Mass balance data available for glaciers in Norway (from Roland 1985)

GLACIER	LAT	LONG	PERIOD	REFERENCE
Ålfotbreen	61 45	5 40	1963-	
Blomsterkardbreen	59 58	6 17	1970-77	Tvede (1972, 1978)
Bondhusbreen	60 02	6 20	1977-81	
Breidablikkbreen & Gråbreen	60 05	6 24	1963-68 1974-75	
Blåbreen & Ruklebreen	60 05	6 26	1963-68	
Midtre Folgefonni	60 08	6 27	1970-71	Tvede (1972)
Vesledalsbreen	61 50	7 16	1967-72	
Nigardsbreen	61 43	7 08	1962-	
Tunbergdalsbreen	61 36	7 03	1966-72	
Store Supphellbre	61 31	6 48	1964-67, 1973-75 1979-82	Orheim (1970) Roland (1985)
Rembesdalsskåkje	60 32	7 22	1963-	
Omnsbreen	60 38	7 29	1966-70	
Storbreen	61 34	8 08	1949-	Liestøl (1967)
Tverråbreen	61 36	8 18	1962-63	Dybwadskog (1965)
Vestre Memurubre	61 32	8 27	1968-72	
Austre Memurubre	61 33	8 30	1968-72	
Hellstugubreen	61 33	8 26	1962-	
Blåbreen	61 33	8 34	1962-63	Klemsdal (1964)
Gråsubreen	61 39	8 36	1962-	
Engabreen	66 39	13 51	1970-	
Høgtuvbreen	66 27	13 39	1971-77	
Trollbergdalsbreen	66 43	14 26	1970-75	
Blåisen	68 20	17 51	1963-68	
Storsteinsfjellbreen	68 06	17 55	1964-68	
Cainhavarre	68 06	18 00	1965-68	
Svartfjelljøkulen	70 14	21 56	1978-79	

the moraines mean that their value to dendroglaciological studies is limited.

The main problem with using glacier length data is that there is always a time lag between a climatic forcing event and a change in glacier length (i.e. the response time). This varies according to local factors and the nature of the forcing event. The response time of Briksdalsbreen (a very steep outlet glacier of the Jostedals icecap) is about 4 years (Liestøl 1967), but is much greater for larger glaciers such as Nigardsbreen (Østrem et al. 1976). Consequently, it may not be possible to establish a direct relationship between glacier frontal positions and tree-ring widths, even if lag periods are used.

The mass balance data collected by the NVE and the Norwegian Polar Institute is much more useful, although the data are available for a limited time period only (Table 1). There is normally a time lag involved between a change in mass balance and a change in the frontal position of an individual glacier, although frontal changes can be modelled from mass balance data with relative ease. Dendroglaciological reconstructions of mass balances have therefore considerable potential in estimating past frontal positions. The longest series is that for Storbreen (34 years) and records extending for 20 years or more are available for five other glaciers (Ålfotbreen, Nigardsbreen, Rembesdalsskåkje, Hellstugubreen and Gråsubreen). The locations of these and other glaciers mentioned in the paper are given in Figure 1. The mass balance data can be divided into the winter balance, which is mainly controlled by the amount of winter precipitation and the summer balance, which is mainly controlled by summer temperature. The relative importance of the summer and winter balances in determining the net annual mass balance varies according to the location of the glacier (Roland 1985). Glaciers in coastal areas are predominantly controlled by the amount of winter accumulation. With increasing continentality the amount of summer ablation becomes important and is the dominant control on the net balance of the most continental glaciers. Therefore, dendroglaciology is more appropriate for the reconstruction of inland glaciers.

The large data base that is available concerning the recent fluctuations of Norwegian glaciers presents many potential lines of research. As already stated, the principal limitation is the shortness of the records. Dendroglaciology provides a potential means of extending the mass balance data back through time, with preliminary studies (Matthews, 1976, 1977a) suggesting that the technique will work.

PAST TREE RING STUDIES IN NORWAY

Norway has a long tradition of dendrochronological work but it does not appear to be widely recognised outside Scandinavia. Many chronologies exist, although a number of problems are associated with their use. The two most important are:

1. Many of the chronologies were prepared prior to 1960 and there is a lack of material from the period 1960-1985. This is important as the calibration of the dendroglaciological work requires chronologies extending to the present time.

2. The methodology employed in the construction of some of the ealier chronologies is unclear. In particular, the method of standardisation

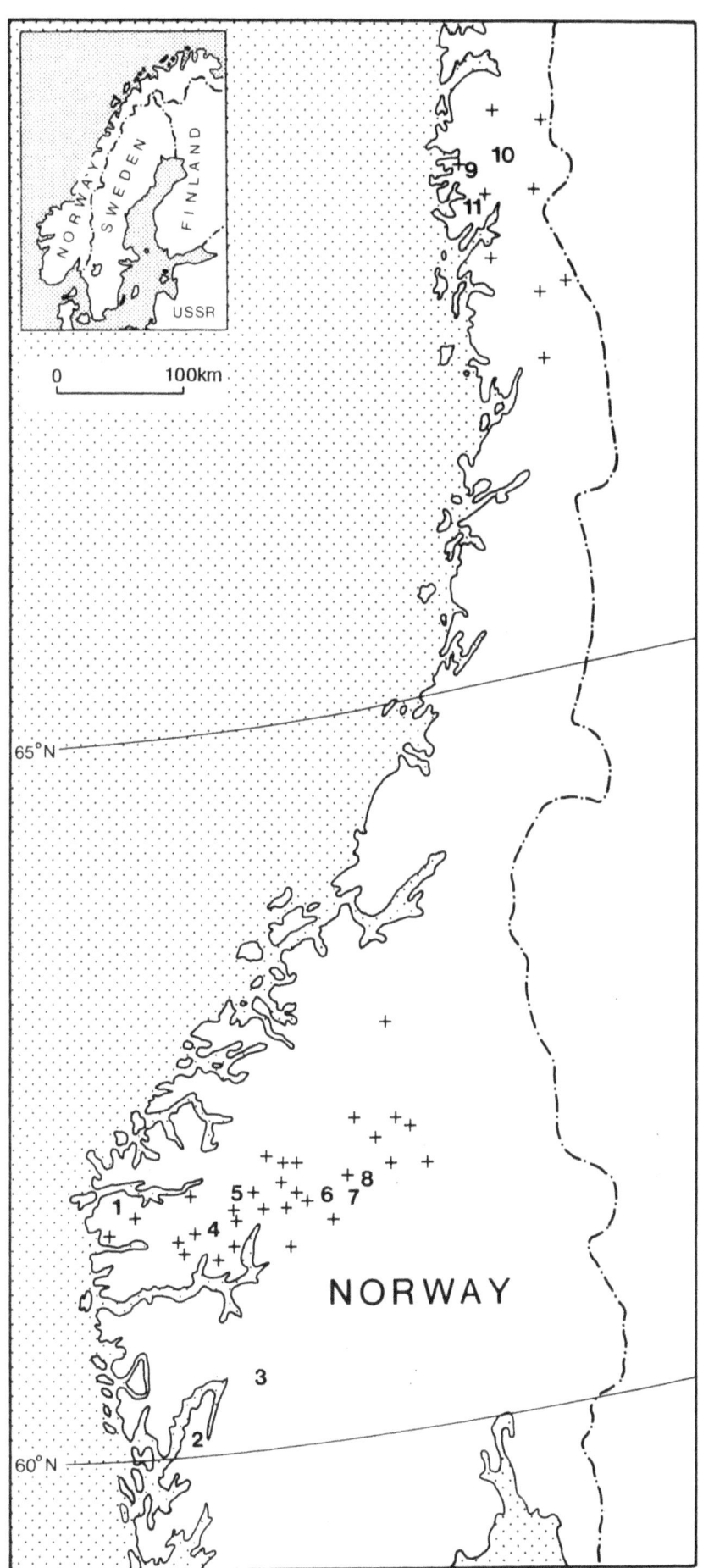

Figure 1. Location of sampling sites (indicated by crosses) and 11 of the glaciers with the longest mass balance records. 1: Ålfotbreen. 2: Folgefonna (Breidalblikkbreen, Blomsterkardbreen). 3: Hardangerjøkulen (Rembesdalssakäkje). 4: Suphellebreen. 5: Nigardsbreen. 6: Storbreen. 7: Hellstugubreen. 8: Gråsubreen. 9: Engabreen. 10: Trollbergdalsbreen. 11: Høgtuvbreen.

employed in some of the chronologies is unsatisfactory, as visual rather than statistical techniques are used.

The majority of chronologies were developed for archaeological dating and are therefore rather unsuitable for dendroclimatic work. They are derived from two main areas, south-east Norway and Trondheim. A small number of chronologies have been obtained from elsewhere. The reliability of the chronologies is unknown although a number of general conclusions can be drawn.

The primary control on the radial growth of both Scots pine (Pinus sylvestris L.) and Norway spruce (Picea abies (L.) Karst.) is summer temperature, which becomes steadily more important towards the tree-line/tree limit. This is the case for Norway (e.g. Eide 1926, Ording 1941), Sweden (e.g. Schweingruber et al. 1979, Aniol and Eckstein 1984) and Finland (e.g. Hustich 1978, Sirén 1961, Mikola 1977). However, summer precipitation may also be important at some sites. For example, Slåstad (1957) argued that in valley bottoms in south-central Norway, precipitation was the primary control on growth, whereas temperature was the dominant control at sites near the tree-line on the valley sides and Kärenlampi (1972) found a high correlation between growth and July precipitation in the Utsjoki valley in northern Finland. On the other hand, several workers have argued that precipitation is unimportant near the northern forest limit (e.g. Hustich and Elfving 1944, Hustich 1945, Sirén 1961), and Svenonius and Olausson (1978) have argued that precipitation has little influence on spruce growing in western Sweden.

In the majority of the studies that have attempted to correlate growth with precipitation, meteorological data have been drawn from stations some distance from the sample site. Precipitation is very variable over any distance and this may be the reason for the variation in the results obtained by different workers.

Whatever the reason, clearly the relationships between climate and tree-growth are rather more complex than simple summer temperature - annual growth correlations. There is a considerable amount of evidence to suggest that spruce and pine behave in different ways and should be dealt with separately (Ording 1941), although Schweingruber and Schär (1976) have not recorded any evidence of this in their densitometric studies. Whether these are due to climatic or other factors (e.g. insect attack) has not yet been evaluated. The annual rings of pine tend to be much more constant over wide areas than spruce (Eklund 1954). Mikola (1950) found that the annual variations in pines were less than in spruce, although long-term variations were clearer in pines. Pines are strongly affected by the temperature of the previous year, whereas spruce rings are hardly affected at all (Eklund 1954, Hoeg 1956, Jonsson 1969). Discrepancies between the two species also occur when the spring has been warm as spruce appears to be able to take advantage of this whereas pine cannot (Ording 1941, Mikola 1950).

At the majority of sites, the annual growth of both spruce and pine is, to a large extent, controlled by summer temperature. The strength of this dependence increases with altitude and towards the north. In some particularly dry situations, summer precipitation may influence growth, and its role may increase in very dry years. The annual growth of pine is partly influenced by the previous years's summer temperature although the extent of this is uncertain. Spruce shows no such autocorrelation effects.

THE PRESENT PROJECT

The current project is aimed principally at developing the basic methodology of dendroglaciology. The sort of problems that are being addressed are:

i) Is there a good relationship between tree growth and summer mass balance at all Norwegian glaciers?

ii) Can tree rings be used to retrodict summer mass balance data?

iii) If a glacier's net annual balance is principally controlled by summer ablation, can tree rings be used to retrodict past annual mass balances?

iv) Can the relationship between periods of glacial advance and depressed tree growth obtained by Matthews for Storbreen be reproduced for other glaciers?

v) How representative are site chronologies in areas such as west Norway which has exceptionally steep environmental gradients?

vi) Are the patterns of tree growth found in north Norway also found in south Norway?

In an attempt to answer some of these problems, an extensive network of sample sites has been established in Norway (Figure 1 and Table 2). The chronologies are concentrated in two main area: Svartisen/Okstindan and an area centred on Jostedalsbreen/Jotunheimen. At each site, two increment cores were taken from each of 15-20 Scots pines and these are being used to develop the basic site chronologies. At the time of writing, only two final chronologies have been developed and these are yet to be examined in detail. The full results of the project will be published in due course.

A number of problems have been encountered during the course of the project. The most obvious is that in western Norway, the forests have had a long history of exploitation. This means that at some sites (e.g. Gaupne) there are very few old trees with the result that some of the chronologies are relatively short. Related to this are the effects of management practices on tree growth. The majority of sites are situated close to the tree-line, where extensive management is unlikely to have occurred. However, the work of Kullman (e.g. Kullman 1983) suggests that the present tree-lines in many parts of Scandinavia are currently below the climatic tree-line due to human interference. Only one plantation site was sampled during the course of the fieldwork: Ottadalen. This is close to some of Slåstad's sites and was sampled in an attempt to reproduce Slåstad's finding that valley-bottom trees were influenced more by precipitation than by temperature.

A further difficulty is that some of the sites may have been affected by atmospheric pollution. One site in particular, Vettifossen, is known to have been affected by fluorine pollution from the aluminium works at Øvre Årdal (Horntvedt 1971). The damage is evident from the large number of dead pines at the site and the suppressed growth of living trees in recent years. A number of Scandinavian studies (e.g. Havas and Huttunen 1972, Soikkeli and Tuovinen 1979, Kvist and Barklund 1984) have demonstrated that

Table 2. Locations of chronologies developed in the current project. Chronology lengths are given to the nearest 50 years.

SITE	LAT	LONG	ALT	LENGTH	NO. OF TREES
00 Leirdalen	61 38	8 10	990	150	15
01 Beiarndalen	66 53	14 45	120	100	12
02 Drivdalen	62 22	9 38	680	200	17
03 Dovrefjell	62 08	9 15	800	300	17
04 Dunderdalen	66 28	15 00	150	250	16
05 Glomdalen	66 27	13 50	98	200	10
06 Halsa	66 45	13 38	80	200	12
07 Boverdalen	61 40	8 10	1000	150	18
08 Luktvatn	66 05	13 40	550	200	15
09 Namdalen	65 18	13 20	359	350	10
10 Ottadalen	61 53	8 20	450	100	17
11 Rossvatnet	65 55	14 25	385	150	13
12 Saltdalen	66 49	15 25	500	350	11
16 Gaupne	61 26	7 15	90	50	13
17 Halsavatnet	61 17	7 05	170	100	10
18 Jostedalen	61 40	7 16	325	100	13
19 Bjorkanosi	61 35	7 17	175	100	13
20 Visdalen	61 43	8 28	900	350	18
22 Meadalen	61 49	8 35	900	100	14
23 Lemonsjoen	61 45	9 05	800	250	14
24 Russdalen	61 34	8 55	970	300	15
25 Vettifossen	61 23	7 58	870	250	16
26 Fortundalen	61 38	7 45	900	200	15
27 Juvasshö	61 41	8 25	900	150	15
28 Kongsfjell	65 53	14 00	310	250	15
29 Svinslåberga	61 49	9 37	900	200	15
30 Annaripig	61 53	9 27	900	200	15
31 Grimsdalen	62 06	9 45	950	200	15
32 Doralen	62 04	9 55	970	350	17
33 Snödola	61 47	10 15	900	300	15
34 Lundadalen	61 48	8 15	1000	100	15
35 Bråtådalen	61 48	7 42	850	350	16
36 Oybergseter	62 02	7 47	900	350	15
37 Fillefjell	61 47	6 31	500	150	15
38 Seldeggja	61 40	5 47	300	300	15
39 Gröfjell	61 39	5 12	250	150	15
40 Rimmane	61 24	6 14	480	150	14
41 Haukedalen	61 25	6 26	380	150	15
42 Rimafjell	61 20	6 27	600	200	16

conifers can be adversely affected by industrial air pollution, although the effects of this seem to be fairly local and the phenomenon is unlikely to have significantly affected growth at the majority of the sample sites. Acid precipitation has also been linked to changes in tree growth in some areas, although its impact in Scandinavia is by no means clear (c.f. Strand 1980, Tveite and Abrahamsen 1980, Jonsson and Sundberg 1982, Arovaara et al. 1984). The present data set, combined with data from other parts of Scandinavia, is currently being used in an attempt to resolve some of the uncertainties surrounding the problem and the results will be incorporated into the dendroglaciological work as appropriate.

The present data set is thus sufficiently large to enable a wide variety of dendroecological problems to be assessed although the initial aims were purely related to dendroglaciology. It is hoped that a full set of recommendations for dendroglaciological procedures will be produced at the close of the project which will enable the technique to be applied in a variety of different glacial environments. As most studies of glacier mass balance are severely limited by the shortness of the records that are available, any means of extending these will be of considerable value. The increasing interest that is being shown in mass balance data (c.f. Kuhn 1984, Reynaud et al. 1984) suggests that the extension of the records will be of use to both climatologists and glaciologists.

Acknowledgements

The research described above was funded by the Natural Environment Research Council. Assistance with the collection of the cores was kindly provided by Catherine Boulton, Nicolas Jeffreys, Joanna Neill, Lucinda Nias, Jonathan Raper, Jennifer Sandall and Louise Taylor.

REFERENCES

Adamenko, V.N., 1963. On the similarity in the growth of trees in northern Scandinavia and in the Polar Ural Mountains. Journal of Glaciology 4: pp. 449-451.

Anderson, J.L. and Sollid, J.L., 1971. Glacial chronology and glacial geomorphology in the marginal zones of the glaciers, Midtdalsbreen and Nigardsbreen, south Norway. Norsk geografisk Tidsskrift 25: pp. 1-38.

Aniol, R.W. and Eckstein, D., 1984. Dendroclimatic studies at the northern timberline. In Mörner, N.-A. and Karlén, W. (eds.), Climatic changes on a yearly to millenial basis, Dordrecht, D. Riedel. pp. 273-280.

Arovaara, H., Hari, P. and Kuusela, K., 1984. Possible effect of changes in atmospheric composition and acid rain on tree growth: an analysis based on the results of Finnish National Forest Inventories. Commentationes Instituti Forestales Fennica 122.

Bircher, W., 1982. Zur Gletscher- und Klimageschichte des Saastales. Glazialmorphologische und dendroklimatologische Untersuchungen. Physische Geographie 9: pp. 1-233.

Bray, J.R. and Struik, G.J., 1963. Forest growth and glacial chronology in Eastern British Columbia and their relation to recent climatic trends. Canadian Journal of Botany 41: pp. 1245-1271.

Denton, G.H. and Karlén, W., 1977. Holocene glacial and tree-line variations in the White River Valley and Skolai Pass, Alaska and Yukon Territory. Quaternary Research 7: pp. 63-111.

Dybwadskog, G., 1965. Tverråbreen. Akkumulasjon og ablasjon i relasjon til data fra fem meteorologiska stasjoner. Unpublished dissertation, University of Oslo.

Eide, E., 1926. Om sommervarmens innflydesle på årringbredden. Meddelelser fra det Norske Skogforsoksvesen, 2 (7): pp. 87-102.

Eklund, B., 1954. Årsringsbreddens klimatiskt betingade variation hos tall och gran inom norra Sverige åren 1900-1944. Meddelelser fran Statens Skogsforskningsinstitutt 44: pp. 1-150.

Faegri, W., 1950. On the variations of western Norwegian glaciers during the last 200 years. Union Géodésique et Géophysique Internationale, Association International d'Hydrologie Scientifique, Assemblée Générale d'Oslo, 19-28 Aout 1948, Tome II: pp. 293-303.

Giardiño, J.R., Schroder, J.F. and Lawson, M.P., 1984. Tree-ring analysis of movement of a rock-glacier complex on Mount Mestas, Colorado, U.S.A. Arctic and Alpine Research 16: pp. 299-309.

Haakenson, N. and Roland, E. 1984. Materialhusholdning, meteorologiske og hydrologiske undersokelser i Norge 1981. Hydrologisk avdeling, Norges vassdrags- og elektrisitetsvesen. Rapport 1/84, pp. 3-59.

Havas, P. and Huttunen, S., 1972. The effect of air pollution on the radial growth of Scots pine (Pinus sylvestris L.). Biological Conservation 4: pp. 361-368.

Heikkinen, O., 1980. Mountain pine radial growth and the forest limit zone in Gudmental, the Swiss Alps. Fennia 158: pp. 1-14.

Heikkinen, O., 1984. Dendrochronological evidence of variations of Coleman Glacier, Mount Baker, Washington. U.S.A. Arctic and Alpine Research 16: pp. 53-64.

Hoeg, O.A., 1956. Growth-ring research in Norway. Tree-ring Bulletin 21: pp. 2-15.

Horntvedt, R., 1971. Fluorskader pa furuskog i Vettismorki. Tidsskrift for Skogbruk 79: pp. 292-301.

Hustich, I., 1945. The radial growth of the pine at the forest limit and its dependence on the climate. Commentationes biologicae. Societas scientiarum fennica 9 (2): pp. 1-30.

Hustich, I., 1978. The growth of Scots pine in northern Lapland, 1928-77. Annales Botanici Fennici 15: pp. 241-52.

Hustich, I. and Elfving, G., 1944. The radial growth variations of forest border pine. Commentationes biologicae. Societas scientiarum fennica 9: pp. 1:18.

Innes, J.L., 1984. Lichenometric dating of moraine ridges in northern Norway: some problems of application. Geografiska Annaler 66A: pp. 341-352.

Jonsson, B., 1969. Studies of variations in the widths of annual rings in Scots pine and Norway spruce due to weather conditions in Sweden. Royal College of Forestry, Research Notes 16, p. 297 (Stockholm).

Jonsson, B. and Sundberg, R., 1982. A study of the effects of air pollution on forest yield. A follow-up of the report of Jonsson and Sundberg 1972 and a new study based on forest types. Avdelningen för skogsuppskattning och skogsindelning 9: pp. 1-61. (Ume).

Kärenlampi, L., 1972. On the relationships of the Scots pine annual ring width and some climatic variables at the Kevo Subarctic Station. Report of the Kevo Subarctic Station 9: pp. 78-81.

Karlén, W., 1976. Lacustrine sediments and treelimit variations as indicators of Holocene climatic fluctuations in Lappland, Northern Sweden. Geografiska Annaler 58A: pp. 1-34.

Karlén, W. 1979. Glacier fluctuations in the Svartisen area, northern Norway. Geografiska Annaler 61A: pp. 11-28.

Klemsdal, T., 1964. En glasiologisk undersokelse i Ost-Jotunheimen. Unpublished dissertation, University of Oslo.

Kuhn, M., 1984. Mass budget imbalances as criterion for a climatic classification of glaciers. Geografiska Annaler 66A: pp. 229-238.

Kullman, L., 1983. Past and present tree-lines of different species in the Handölan valley, central Sweden. In Morisset, P. and Payette, S., eds., Tree-line Ecology. Proceedings of the Northern Quebec Tree-line Conference. Quebec, Centre d'études nordiques Université Laval. Nordicana 47: pp. 25-45.

Kvist, K. and Barklund, P., 1984. Luftverschmutzungsprobleme in schwedischen Wäldern. Fortwissenschaftliches Centralblatt 103: pp. 74-82.

LaMarche, V.C. and Fritts, H.C., 1971. Tree rings, glacial advance, and climate in the Alps. Zeitschrift für Gletscherkunde und Glazialgeologie 7: pp. 125-131.

Liestøl, O., 1967. Storbreen glacier in Jotunheimen, Norway. Norsk Polarinstitutt Skrifter 141: pp. 1-63.

Lovelius, N.V., 1972. Reconstruction of the course of meteorological processes on the basis of the annual tree rings along the northern and altitudinal forest boundaries. In Wielgolaski, F.E. and Rosswall, R. (eds.), Tundra Biome. Stockholm, Swedish I.B.P. Committee, Wenner-Gren Centre, pp. 248-268.

Matthews, J.A., 1974. Families of lichenometric dating curves from the Storbreen gletschervorfeld, Jotunheimen, Norway. Norsk geografisk Tidsskrift 28: pp. 215-235.

Matthews, J.A., 1975. Experiments on the reproducibility and reliability of lichenometric dates, Storbreen gletschervorfeld, Jotunheimen, Norway. Norsk geografisk Tidsskrift 29: pp. 243-245.

Matthews, J.A., 1977a. Glacier and climatic fluctuations inferred from tree-growth variations over the last 250 years, central southern Norway. Boreas 6: pp. 1-24.

Matthews, J.A., 1977b. A lichenometric test of the 1750 end-moraine hypothesis: Storbreen gletschervorfeld, southern Norway. Norsk geografisk Tidsskrift 31: pp. 129-136.

Mikola, P., 1950. On variations in tree growth and their significance to growth studies. Commentationes Instituti Forestales Fennica 38(5): pp. 126-131.

Mikola, P. 1977. Consequences of climatic fluctuation in forestry. Fennia 150: pp. 39-43.

Ording, A., 1941. Årringanalyser pa gran og furu. Meddelelser fra det Norske Skogsforsoksvesen 7: pp. 105-354.

Orheim, O., 1970. Glaciological investigations of Store Supphellebre, West Norway. Norsk Polarinstitutt Skrifter 151: pp. 1-48.

Ostrem, G., Liestøl, O. and Wold, B., 1976. Glaciological investigations at Nigardsbreen, Norway. Norsk geografisk Tidsskrift 30: pp. 187-209.

Paterson, W.S.B., 1981. The Physics of Glaciers. Oxford, Pergamon. p. 380.

Renner, F., 1982. Beiträge zur Gletschergeschichte des Gotthardgebietes und Dendroklimatologische Analysen an Fossilen Hölzern. Physische Geographie 8: pp. 1-182.

Reynaud, L., Vallon, M., Martin, S. and Letreguilly, A., 1984. Spatio-temporal distribution of the glacial mass balance in the Alpine, Scandinavian and Tien Shan areas. Geografiska Annaler 66A: pp. 239-247.

Röthlisberger, H., 1976. Gletscher- und Klimaschwankungen im Raum Zermatt, Ferpécle und Arolla. In: 8000 Jahre Walliser Gletschergeschichte. Ein Beitrag zur Erforschung des Klimaverlaufs in der Nacheiszeit. Die Alpen No. 3/4: pp. 59-152.

Roland, E. 1985. Materialhusholdningen, meteorologiske og hydrologiske undersokelser ved utvalgte breer. In Roland, E. and Haakensen, N. (eds.), Glasiologiske undersokelser i Norge 1982, Oslo, Norges Vassdrags- og Elektrisitetsvesen, pp. 4-43.

Schneebeli, W., 1976. Untersuchungen von Gletscherschwankungen im Val de Bagnes. In: 8000 Jahre Walliser Gletschergeschichte. Ein Beitrag zur Erforschung des Klimaverlaufs in der Nacheiszeit. Die Alpen No. 3/4: pp. 5-57.

Schweingruber, F.H., Bräker, O.U. and Schär, E., 1979. Dendroclimatic studies on conifers from central Europe and Great Britain. Boreas 8: pp. 427-452.

Schweingruber, F.H. and Schär, E., 1976. Röntgenuntersuchungen an Jahrringen. Neue Zürcher Zeitung 180, 4. Aug. 1976. p. 33.

Sigafoos, R.S. and Hendricks, E.L., 1961. Botanical evidence of the modern history of Nisqually Glacier, Washington. United States Geological Survey Professional Paper 387-A: pp. 1-20.

Sirén, G., 1961. Skogsgränstallen som indikator för klimatfluktuationerna i norra fennoskandien under historisk tid. Commentationes Instituti Forestales Fennica 54: pp. 1-66.

Slåstad, T., 1957. Tree-ring analyses in Gudbrandsdalen. Meddelelser fra det Norske Skogforsoksvesen 48: pp. 571-620.

Soikkeli, S. and Tuovinen, T., 1979. Damage in mesophyll ultrastructure of needles of Norway spruce in two industrial environments in central Finland. Annales Botanici Fennici 16: pp. 50-64.

Svenonius, B. and Olausson, E., 1978. Ring widths of trees, solar activity and weather conditions in Sweden in the period 1756-1975. Geologiska Föreningens i Stockholm, Förhandliger 100: pp. 95-100.

Tvede, A.M., 1972. En glasio-klimatisk undersokelse av Folgefonni. Unpublished dissertation, University of Oslo.

Tvede, A.M., 1978. Blomsterkardbreen. In Wold, B. and Haakensen, N., Glasiologiske undersokelser i Norge 1977. Hydrologisk avdeling, Norges vassdrags- og elektristetsvesen, Report 4/79, pp. 11-12.

Tveite, B. and Abrahamsen, G., 1980. Effects of artificial acid rain on the growth and nutrient status of trees. In Hutchinson, T.C. and Havas, M., (eds.), <u>Effects of acid precipitation on terrestrial ecosystems</u>, Plenum Press, pp. 305-318.

Mediaeval Dendrochronology in Exeter and its Environs

J. Hillam and C. M. Mills
Sheffield Dendrochronology Laboratory
Department of Archaeology and Prehistory
University of Sheffield
Sheffield S10 2TN

ABSTRACT

A new research project on the timbers from Exeter Cathedral roof began in January 1985. Previous research in south-west England is reviewed, and the aims and potential of the current research are discussed in terms of its importance to general dendrochronological research and to the history and archaeology of south-west England.

INTRODUCTION

A series of mediaeval roofs are preserved at Exeter Cathedral, and with 160 trusses above the high vaults alone, these provide a wealth of material suitable for dendrochronological research. The existence of detailed contemporary documentation, known as the Fabric Rolls (Erskine 1981, 1983), which relates to the accounts for the building and maintenance of the Cathedral, enhances the research potential of this material. As well as examining the aims of the current research, this paper also discusses the project in the light of previous work in the area.

EXISTING RESEARCH

Previous research in the Exeter region has produced several dated oak tree-ring chronologies (Figure 1). With the increase in rescue archaeology in the 1970's, several large archaeological excavations were carried out in Exeter's city centre prior to redevelopment, and these yielded many tree-ring samples. Waterlogged timbers from Trichay Street were used to provide a chronology which dated to AD 816-1216 (Hillam 1984). Excavations at nearby Goldsmith Street produced four groups of timbers, one of which produced a chronology for the period AD 775-1022. The other three groups, two of oak (Quercus spp) and the other of beech (Fagus sylvatica), remain undated (Morgan 1984). A single timber from the medieval Exe Bridge in Exeter was also dated. Its ring sequence covers the period AD 799-941, and matches well with the Trichay chronology (Hillam 1980).

Tree-ring work on timber buildings has also been successful. The first Devon building to be dated was Reed Farm, Christow, which produced a sequence from AD 1460 to 1527 (Morgan 1980). A 166-year chronology was constructed from two timbers from the Guildhall, Exeter, and dated to AD 1424-1589 (Bridge 1983). Two oak boards from the Bishop's Throne, Exeter Cathedral, had ring sequences which, when combined, gave a 183-year chronology. Since there was no sapwood on either sample, a precise felling date could not be estimated, but the last ring of the sequence was AD 1284.

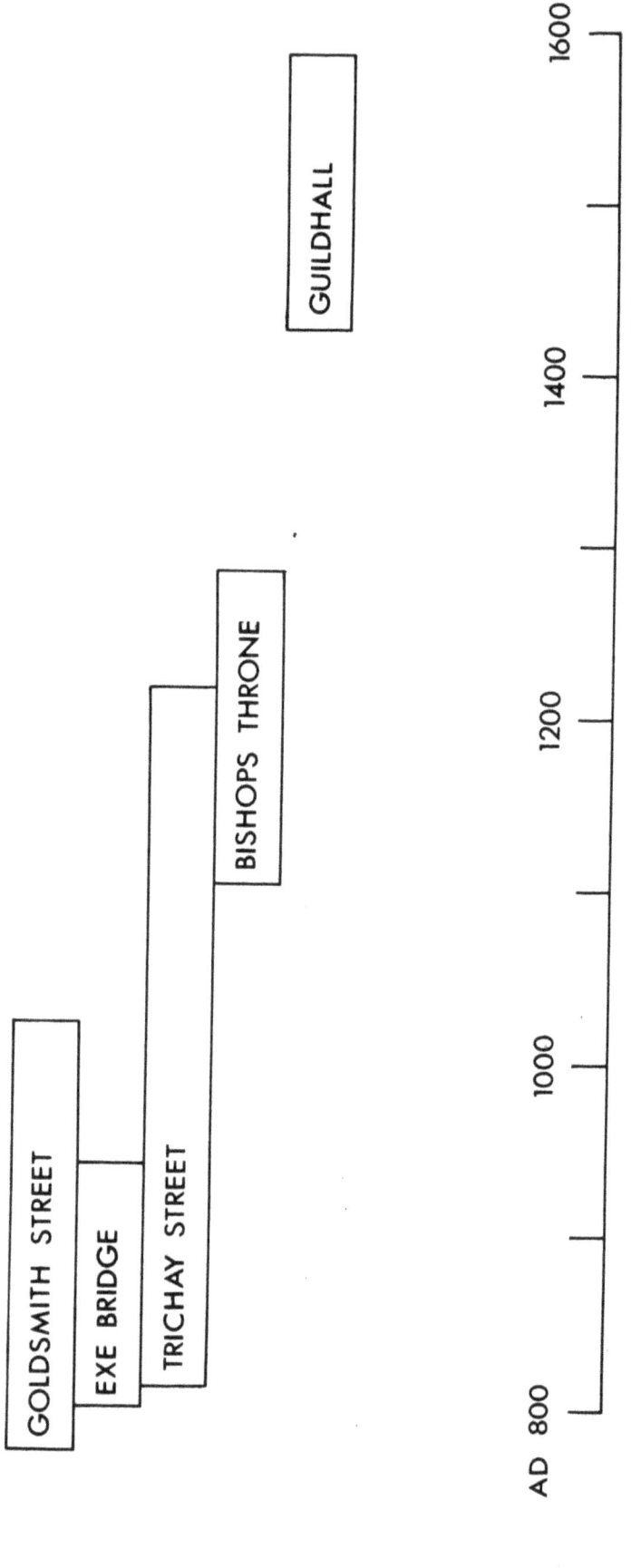

Figure 1. Tree-ring sequences from Exeter. Bar diagram showing the temporal relationship of dated chronologies. Each bar represents one site (see text for references).

This, together with reference in the Fabric Rolls to the import of oak boards from the Continent, suggests an early 14th century date for the timbers (Bridge 1983 p. 110).

Other attempts at dating timbers from this region have been less successful. Work on material from Great Moor Farm, a 16th century cruck building near Exeter (Hillam unpubl), exemplified the recurrent problems provided by the use of young, wide-ringed trees in the construction of timber-framed buildings. At Goldsmith Street, the low mean sensitivity of some timbers prevented the dating of the second oak chronology (Morgan 1984). Bridge (1983 p. 124) was unable to date wood samples from various Exeter buildings, whilst late- and post-mediaeval timbers from Trichay Street were also found to be undatable. Bridge (1983 p. 127) concluded that lack of dating is due to complacent ring patterns from trees with no more than 60-70 wide rings, and the inability to produce site chronologies.

CURRENT RESEARCH

It is hoped that current research on material from Exeter Cathedral, and the examination of the Fabric Rolls, will determine more precisely how timber provenance affects the crossmatching procedure. The contemporary documentation provides the opportunity to ascertain provenance, and to investigate its effect upon the crossmatching and dating of samples from the Cathedral, as well as those from other Devon buildings. It will also provide a chance to study some aspects of the mediaeval timber trade which are relevant to dendrochronological interpetation, particularly stockpiling and seasoning practices.

The use of wood from different sources is one of the most likely reasons for the lack of crossmatching between timbers from the same site (Hillam 1985), particularly if some of the timbers have been imported from the Continent (Baillie et al. 1985). Exeter's coastal location and role as port, plus its proximity to trading centres such as Bristol and Plymouth, may have favoured the use of imported timber. The work on the Cathedral timbers might solve the kind of problem encountered at Trichay Street where one group of timbers gave an exceptionally high correlation with the reference chronology for Dublin (Baillie 1977). Debate continues as to whether the Trichay timbers were imported from Ireland, although archaeological evidence tends to favour a local origin.

A comprehensive sampling programme will also provide material with which to date the different phases of roof-construction at Exeter Cathedral. The vast number of timbers available for sampling should help to overcome any problems caused by short, complacent ring patterns, and the data they produce will be relevant to more general dendrochronological research, such as the statistical treatment of tree-ring data. Work on this aspect of the research is being carried out in the Department of Probability and Statistics at Sheffield University. Here the emphasis is placed upon the dating of material from buildings or archaeological sites where short ring sequences are common, rather than on the dating of long sequences used for chronology building.

Finally the work should result in the production of a long tree-ring chronology for south-west England which, together with the new approach to short sequences and complacent timbers mentioned earlier and discussed in Hillam et al. (this volume), should provide a dating framework for any mediaeval timbers from the region.

Acknowledgements

The Sheffield Dendrochronology Laboratory is financed by the Historic Buildings and Monuments Commission for England. C. Mills is funded by a SERC/Case award, which is sponsored by the Archaeology Unit at the Royal Albert Museum, Exeter, under the supervision of John Allan.

REFERENCES

Baillie, M.G.L. 1977. Dublin Medieval Dendrochronology. Tree Ring Bulletin 37 pp. 13-20.

Baillie, M.G.L., Hillam, J., Briffa, K.R. & Brown, D.M. 1985. Re-dating the English art-historical tree-ring chronologies. Nature 315 pp. 317-319.

Bridge, M. 1983. The use of Tree-Ring Widths as a Means of Dating Timbers from Historical Sites. PhD thesis, C.N.A.A. (Portsmouth Polytechnic).

Erskine, A.M. (ed & translator) 1981. The Accounts of the Fabric of Exeter Cathedral, 1279-1353. Part 1: 1279-1326. Devon and Cornwall Record Society, New Series, Vol. 24.

Erskine, A.M. (ed & translator) 1983. The Accounts of the Fabric of Exeter Cathedral, 1279-1353. Part 2: 1328-1353. Devon and Cornwall Record Society, New Series, Vol. 26.

Hillam, J. 1980. A medieval oak tree-ring chronology from south-west England. Tree Ring Bulletin 40 pp. 13-22.

Hillam, J. 1984. Tree-ring analysis of Trichay Street timber, Exeter. In: Medieval and Post-Medieval Finds from Exeter, 1971-1980 (ed. Allan, J.P.) Exeter Archaeological Reports 3 pp. 315-318.

Hillam, J. 1985. Recent tree-ring work in Sheffield. Current Archaeology 9(1) pp. 21-26.

Hillam, J., Morgan, R.A. and Tyers, I. 1987. (this volume) Sapwood Estimates and the Dating of Short Ring Seqeunces. pp. 165-185.

Morgan, R.A. 1980. Tree-ring dates for buildings. Vernacular Architecture 11 p. 22 (List 1).

Morgan, R.A. 1984. Tree-ring analysis of Goldsmith Street timber, Exeter. In: Medieval and Post-Medieval Finds from Exeter, 1971-1980 (ed Allan, J.P.) Exeter Archaeological Reports 3 pp. 318-322.

A 700 Year Dating Chronology for Northern France

J.R. Pilcher

Palaeoecology Centre
Queen's University of Belfast
Northern Ireland

ABSTRACT

An oak tree-ring chronology spanning 1274-1979 has been completed for Northern France using timbers from living trees and historic buildings mainly in the Loire Valley. The process of chronology construction in a new area is illustrated and the chronology is presented to assist with dating in the area and to form the basis for the construction of other regional chronologies in France.

INTRODUCTION

This paper reports the completion of a dendrochronological sequence for mediaeval and post-mediaeval buildings and living trees in northern France. As is usual in the early stages of a dendrochronological project the intention was to build the dating framework rather than to try to solve major architectural dating problems. Now that the chronology has been established the method is available for the dating of structures built of oak in the last 700 years in northern France. The tree-ring chronology is presented here with some details of the individual sites studied.

PREVIOUS DENDROCHRONOLOGY IN FRANCE

So far there seems to have been no serious attempt at producing a continuous tree-ring chronology for northern France. In the last 20 years major centres of tree-ring studies have developed in Germany and in Ireland. Long oak chronologies tied to living trees have been produced for northern Ireland (Baillie, 1977a), Southern Scotland (Baillie, 1977b), Northern Germany (Hollstein, 1980), Southern and Central Germany (Becker and Delorme, 1978), Denmark (Bartholin, 1975) and the Netherlands (Eckstein, 1978). Both Northern Ireland and Germany now have chronologies more than 5000 years in length (Pilcher et al., 1984). So far no such chronologies have been published for England or for France. Extensive dendroclimatological studies have been carried out in the South of France and in the Alps by workers at the University of Marseilles (e.g. Serre et al., 1966; Serre, 1978) and in Belgium by A. Munaut (1978). Dr. Siebenlist-Kerner of the University of Hamburg has produced a working chronology for oaks in Northern France for the period 1270-1490 A.D. This unpublished chronology was kindly made available to the present author. It is dated by reference to the north Germany chronologies rather than being tied to living tree chronologies. The Centre Technique du Bois in Paris has produced a living tree chronology of 362 years from north of Paris (Trenard and Duchateau, 1985). Leboutet and Colardelle (1983) studied waterlogged timbers from a mediaeval site at Charavines dating to 968 A.D.

METHODS

Sampling

Samples of living trees were obtained using a 40 cm Swedish increment corer giving a core of about 5mm diameter. A large part of the work in historic buildings involved sampling timbers that had been removed from the structure during repairs or alterations. These timbers could most easily be sampled by hand-saw. Slices of timbers obtained in this way are preferable to cores as there is a larger part of the circumference of the tree available for study. At Loches and Langeais, where loose timbers were not available, cores were extracted from structural timbers using a Henson dry wood corer powered by a hand-held electric drill. For future work on buildings where specific phases or even specific timbers have to be dated more use will have to made of the corer and a portable generator.

Sample preparation and measurement

Cores and slices were air dried, polished and then measured on a Henson Incremental Measurement Machine attached to an Apple computer. As well as logging the data measurement from the Henson machine the Apple microcomputer was used for all the cross-dating tests, formation of site master chronologies and the construction of the final 700-year chronology.

Cross-dating

Most within-site cross-dating was established visually using the plotted ring-widths. The positions selected by eye were then tested for significance of cross correlation using the computer. At each site those trees that could be easily cross-dated were combined to form a site chronology in which each annual value represented the mean ring width for all the trees with rings belonging to that year. Of the 110 samples of non-living timbers collected almost all those with greater than 100 annual rings were cross-dated and about half of those with 60 or more rings. Once cross-dating within the site was established, cross-dating with other chronologies was sought. Initially only the Siebenlist-Kerner chronology and the long living tree chronology from Fontainebleau were available, but as more sites were processed, so more chronologies were available for cross-dating.

Once all the sites had been studied, those that had produced chronologies were combined into a provisional continuous chronology of 700 years. Undated sites were then tested against this long sequence and further cross-dating established. Finally all the dated sites were combined into the version of the Northern France chronology presented here. It should be noted that the chronology presented here could not have been assembled without the existence of the Siebenlist-Kerner chronology and thus the German chronologies by which it is dated. Some of the overlaps between sequences that comprise the France chronology are too short to have been reliably established without reference to other chronologies. This version of the Northern France chronology does not include the chronology of Dr. Siebenlist-Kerner, and cross-dates with it with a t-value of 6.6. It also cross-dates directly with the north German chronology.

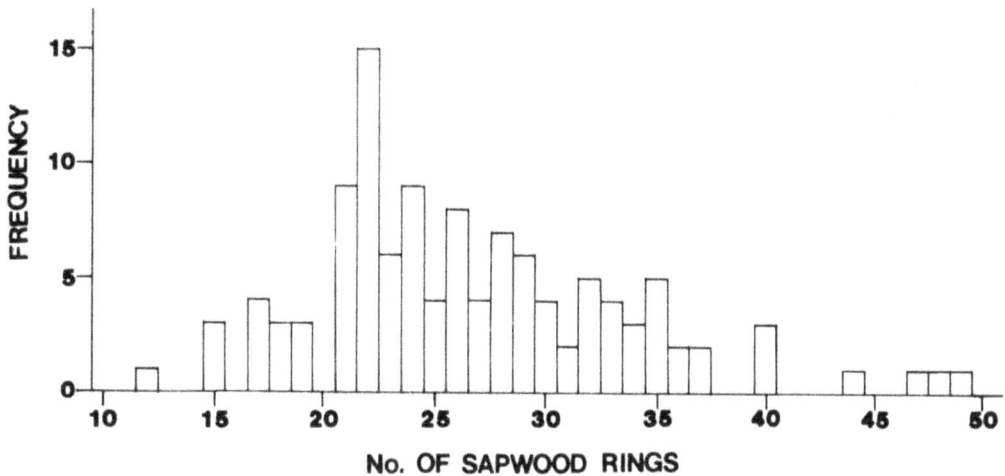

Figure 1. Distribution of the number of sapwood rings in 118 samples of Quercus robur and Q. petraea from northern France.

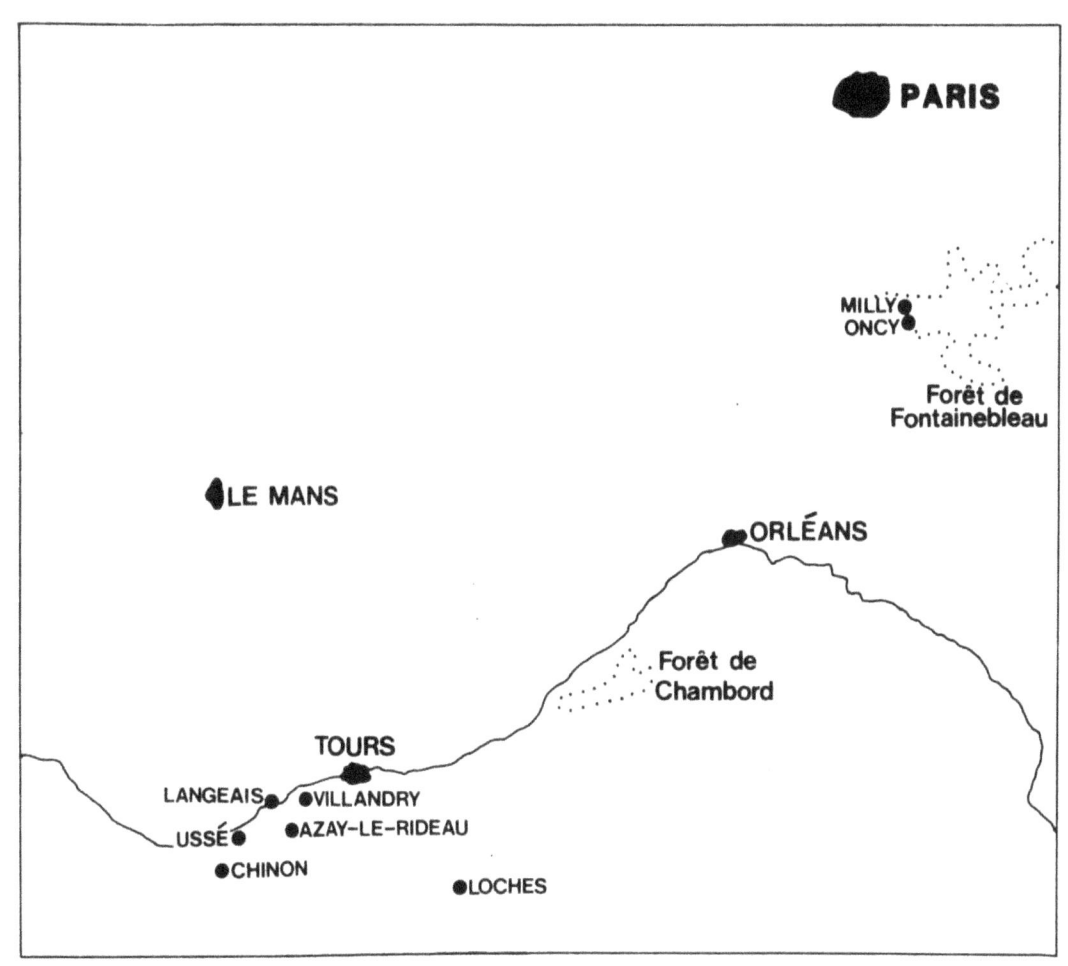

Figure 2. Locations of the forests and buildings studied.

Sapwood Estimates

As it is often necessary to estimate the amount of sapwood missing from the outside of a sample of building timber there has been considerable research into the number of sapwood rings characteristic of oak. This work has been recently summarised by Hughes et al., (1981) and is further discussed by Hillam et al. in this volume. Because of the range of values found in different areas local estimates were made in northern France. By counting the sapwood rings on all the modern cores used for the six modern chronologies developed in northern France, a sample of 118 trees could be used.

Figure 1 shows the distribution of the number of sapwood rings for the modern trees. As also found by Hughes et al. 1981, the distribution is skewed. The mean and standard deviation were calculated on the logarithms of the sapwood numbers, giving a mean of 25.3 and a 95% range of 15 to 43. This compares with the mean of 24.7 and range of 13.7 to 44.6 found by Hughes for British Isles sites. Hughes suggests that as cores from modern trees are taken at about breast height the number of sapwood rings should be increased by a factor of 1.217 to be applicable to the rest of the tree. Accordingly a value of 31 for the mean and 18-52 for the range is suggested on the basis of the northern France trees. It is important to notice that while the range covering 95% of trees is 18 to 52, it is possible for the occasional tree to have as many as 90 sapwood rings or as few as 5. The problems of correctly estimating the number of sapwood rings underlines the importance of searching for the presence of total sapwood when sampling a building for dating purposes. Even one or two samples with total sapwood in a set of samples will greatly improve the reliability of the final age estimate.

THE SITES

The distribution of sites sampled is shown in Figure 2. The two living tree chronologies used here form part of a series produced for a climate reconstruction project and will be published elsewhere. They are both from major areas of oak forest under the management of the Office National des Forêts, and were sampled with the help and permission of the local O.N.F. staff. The very long-lived trees from Fontainebleau are part of a nature reserve area in the Fontainebleau forest, which was protected in former times as a royal hunting forest. The nature reserve represents only a small remnant. The trees were considerably older than the 1531 A.D. starting date of the chronology because our 40 cm long corers could not reach the centre of trees greater than 1 m in diameter. The estimated age of some of the trees is well in excess of 500 years. At both sites about 20 trees were sampled, taking one core per tree from trees showing straight growth.

The historic buildings were chosen to give a reasonable age range and partially decided by where access could be organised within the short time available for field work in France. A selection of the sites will be described to illustrate the chronology construction.

1: Loches

The château of Loches, 40 km SE of Tours, is one of the earliest

TABLE 1

	Q4066	Q4064	Q4069	Q4068	Q4072	Q4073	Q4065	Q4070	Q4071
Q4066	-	5.0	4.1	5.2	4.9	5.5	3.0	6.4	3.8
Q4064	5.0	-	4.1	7.3	5.4	3.9	3.6	2.5	3.7
Q4069	4.1	4.1	-	7.1	8.4	7.6	2.7	4.3	4.8
Q4068	5.2	7.3	7.1	-	11.3	6.2	3.5	4.5	5.5
Q4072	4.9	5.4	8.4	11.3	-	6.9	4.0	5.1	3.5
Q4073	5.5	3.9	7.6	6.2	6.9	-	5.5	6.2	3.5
Q4065	3.0	3.6	2.7	3.5	4.0	5.5	-	2.9	1.2
Q4070	6.4	2.5	4.3	4.5	5.1	6.2	2.9	-	3.4
Q4071	3.8	3.7	4.8	5.5	3.5	3.5	1.2	3.4	-
Mean	4.7	4.4	5.4	6.3	6.2	5.6	3.3	4.4	3.7

Matrix of t-values for the 9 samples from Loches, phase 1.

TABLE 2

	Q4036	Q4033	Q4027	Q4034	Q4037	Q4029	Q4026	Q4038
Q4036	-	13.0	7.5	9.5	7.1	8.5	6.3	6.1
Q4033	13.0	-	7.6	8.1	7.1	7.8	5.3	6.2
Q4027	7.5	7.6	-	7.6	5.0	8.1	6.7	2.5
Q4034	9.5	8.1	7.6	-	6.4	8.6	8.6	3.4
Q4037	7.1	7.1	5.0	6.4	-	4.4	3.2	4.4
Q4029	8.5	7.8	8.1	8.6	4.4	-	6.9	3.3
Q4026	6.3	5.3	6.7	8.6	3.2	6.9	-	4.4
Q4038	6.1	6.2	2.5	3.4	4.4	3.3	4.4	-
Mean	8.3	7.9	6.4	7.5	5.3	6.8	5.9	4.3

Matrix of t-values for the trees of the second phase at Ussé

TABLE 3

	Q4023B	Q4015	Q4013	Q4016	Q4009	Q4022	Q4011	Q4014	Q4012B	Q4024B	Q4010	Q4018	Q4025
Q4023B	-	3.9	2.4	3.7	3.6	3.8	2.1	6.0	9.4	5.0	3.6	4.4	8.1
Q4015	3.9	-	4.1	4.0	6.5	6.3	2.5	3.6	4.6	5.0	3.3	4.0	5.9
Q4013	2.4	4.1	-	3.7	7.6	6.3	5.0	1.6	3.6	4.2	2.8	4.3	3.8
Q4016	3.7	4.0	3.7	-	3.2	4.4	2.6	4.2	2.8	3.7	1.6	4.1	2.4
Q4009	3.6	6.5	7.6	3.2	-	5.1	3.0	1.9	4.7	5.7	3.6	4.1	6.0
Q4022	3.8	6.3	6.3	4.4	5.1	-	3.2	2.1	4.1	5.2	4.1	4.8	5.9
Q4011	2.1	2.5	5.0	2.6	3.0	3.2	-	4.5	2.8	3.3	3.6	5.7	3.1
Q4014	6.0	3.6	1.6	4.2	1.9	2.1	4.5	-	2.9	2.7	3.2	4.8	3.9
Q4012B	9.4	4.6	3.6	2.8	4.7	4.1	2.8	2.9	-	4.2	3.9	5.1	10.8
Q4024B	5.0	5.0	4.2	3.7	5.7	5.2	3.3	2.7	4.2	-	3.4	4.5	4.2
Q4010	3.6	3.3	2.8	1.6	3.6	4.1	3.6	3.2	3.9	3.4	-	5.0	3.8
Q4018	4.4	4.0	4.3	4.1	4.1	4.8	5.7	4.8	5.1	4.5	5.0	-	4.8
Q4025	8.1	5.9	3.8	2.4	6.0	5.9	3.1	3.9	10.8	4.2	3.8	4.8	-
Mean	4.7	4.5	4.1	3.4	4.6	4.6	3.4	3.4	4.9	4.3	3.5	4.6	5.2

Matrix of t-values for the trees of the first phase at Ussé

fortifications in the area that still retains any original timbers. The only building sampled at Loches was the Logis Royaux, a two phase building. Sampling was carried out in the attics where the carpentry styles of the two phases were clearly distinct. Most of the samples were from the older section and included five cores from the carved tie beams which must have been original.

All but one of the early phase samples cross-dated and this had only 60 rings. The internal cross-dating of the 9 samples is summarised in the matrix of t-values given in Table 1, which were calculated as described by Baillie and Pilcher, 1973. The relative positions of the samples is shown in Figure 3. Only two samples show clear indications of having sapwood, one ending in 1334, the other in 1350, which would suggest felling dates in the range 1352-1386 and 1368-1402 using the sapwood estimate described above. A construction date in the 1383-1402 range seems most likely. The Phase One timbers were combined into a chronology of 92 years which cross-dates with the Siebenlist-Kerner chronology with a t-value of 5.8.

2: Tours

A number of old oak timbers were found in a timber yard in the village of Azay-le-Rideau. The workshop specialised in the restoration of old buildings and in work with oak timbers. Of the timbers sampled, 11 were large square beams, described as being from a building in the old quarter of Tours. The timber was of excellent quality for dendrochronology and was sampled in order to provide a building block in the chronology rather than for any slight value the dating of unprovenanced timbers might have.

All 11 timbers cross-dated well with each other. Four samples had sapwood; one had the bark present and the group of 11 samples appears to represent a single construction phase. The chronology formed from the 11 samples is dated against the Siebenlist-Kerner chronology with a t-value of 5.9 and against the Ussé chronology with a t-value of 4.0. These both establish the end date of the Tours chronology to be 1502. The completed Tours chronology spans 1328-1502 and forms a useful contribution to the Northern France chronology. This site illustrates the ease with which dendrochronology can be carried out when renovation (or in this case demolition!) has rendered the timbers accessible for sampling without restriction.

3: Langeais

This imposing château was built by Jean Bourre between 1465 and 1469. It has remained essentially unmodified up to the present day with no major rebuilding or additions and is now owned by the Institut de France who kindly gave permission for sampling. Four cores were extracted from rafters in the attics. Two of the cores had less than 50 rings and were therefore not usable, but the other two cross-dated with each other and with the Northern France chronology. The outer rings of the two samples formed in 1413 and 1431. Neither showed signs of sapwood and are thus likely to have been felled sometime after 1449. Felling during the recorded construction period of 1465-9 is entirely possible. Had it been possible to sample some of the large timbers with sapwood in the towers a more precise felling date would have been obtained. The t-value for the cross-dating of the mean of the two timbers with the provisional France chronology

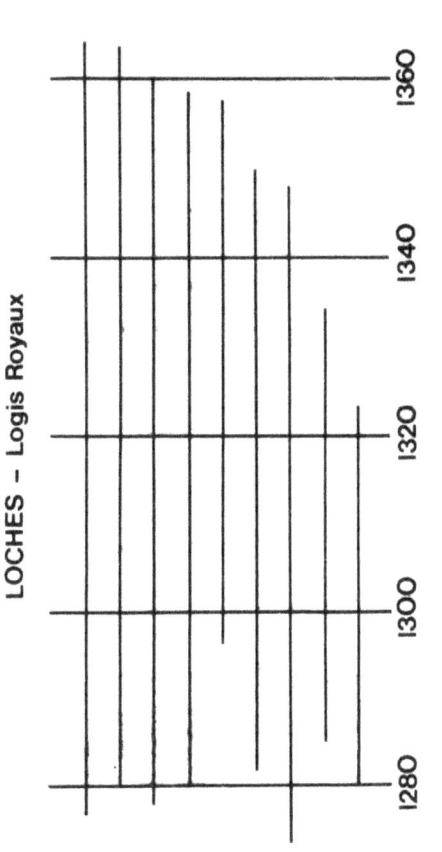

Figure 3. Relative positions of the samples from Phase 1 at Loches.

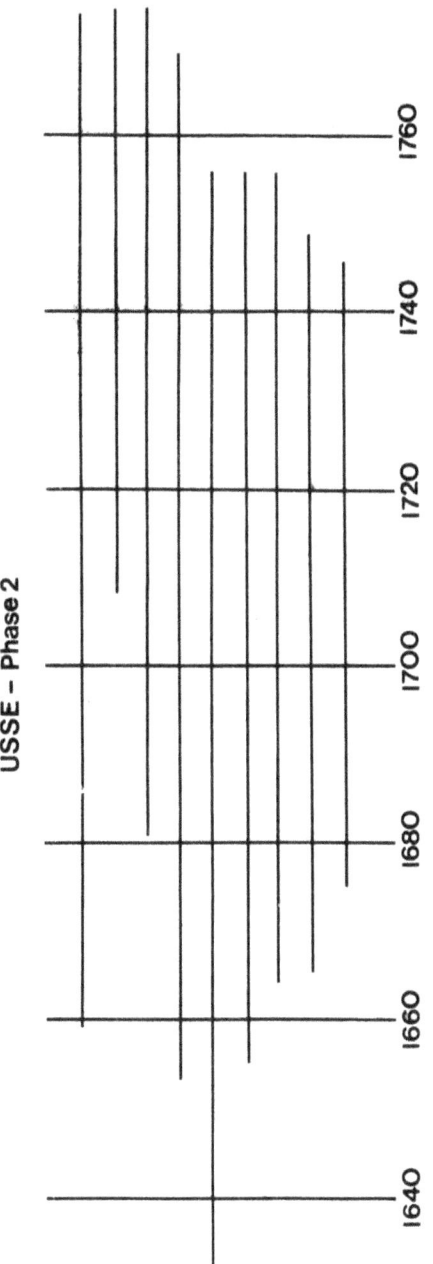

Figure 4. Relative positions of the Phase 2 samples from Ussé. The three top samples had total sapwood and ended in 1774. The pairs of vertical lines mark the heartwood-sapwood transitions.

was 7.6.

4: Ussé

In comparison with Langeais, Ussé is a far more complex structure that has been continuously improved, altered and added to during its history. The château is now the family home of the Comte de Blacas and with his help it was possible to collect 34 samples from the attics of the château. Almost all the samples taken were from loose timbers that could be sawn by hand. A great many were sections of rafters that had been cut out to make way for the insertion of chimneys. In addition to these timbers some purlins with long tapering overlapped joints were sampled from the ends of the tapered portions. Two main groups of samples were taken, the first from the main E-W gallery at the rear of the building and the other from the central N-S gallery.

Most of the samples of group 2 from the N-S gallery dated to a single construction phase. The t-value matrix is given in Table 2 for the nine samples cross-matched. Two undated samples from the N-S gallery cross-dated against each other but not against the other samples. Figure 4 shows the relative positions of the nine samples and the presence of sapwood. The two samples with bark have the same outer year (1774) and those with some sapwood are compatible with this felling date. If it is is assumed that the trees were felled in the winter of 1774/5, construction during 1775 is most likely. The dating of the Ussé chronology was achieved by reference to the Fontainebleau modern tree chronology (t = 6.5) and also by reference to the North German chronology. This shows the advantage of a very long living tree chronology when building a dating chronology in a new area.

Of the 17 samples from the first phase, three were too short to date. The remaining 14 cross-dated with each other as shown in the matrix of t-values given in Table 3. One tree showed poor correlation values (Q4017) although it was satisfactorily cross-dated visually; it was not included in the site chronology. None of the 14 samples had complete sapwood and only two had sapwood-heartwood transitions, which occurred in the years 1479 and 1487. Using the sapwood estimates given above, felling dates in the range of 1497-1531 and 1505-1539 are likely.

The Ussé Phase One chronology spans 1351-1494 and cross-dates with the Tours chronology with a t-value of 4.0 and with the Siebenlist-Kerner chronology with a t-value of 4.4.

5: Milly-la-Forêt

The fine timbered market hall at Milly-la-Forêt is recorded as being constructed in 1479. Timbers from it had been studied by Dr. Siebenlist-Kerner and contributed to her 1270-1490 A.D. chronology. The present author sought permission from the mayor of Milly to take further cores from the hall to improve the chronology. While permission for this was refused, attention was drawn to a large timber converted into park benches in the garden of the Chapelle-Saint-Blaise in Milly. This timber is reputed to have been removed from the hall some years ago when it was damaged by a lorry. Permission was given for the extraction of four cores from the timber which cross-dated to form a chronology of 161 years. Attempts to

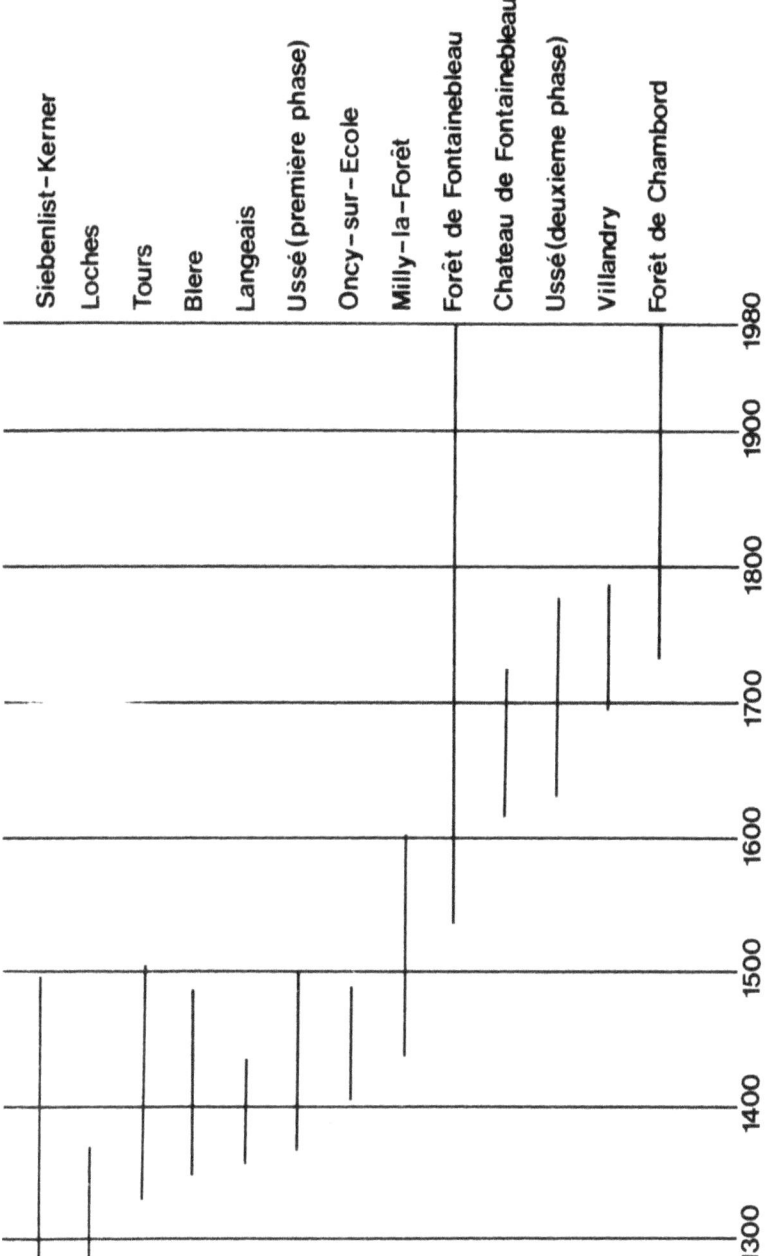

Figure 5. Distribution in time of the site units that make up the Northern France chronology together with the Siebenlist-Kerner chronology.

date this at the expected position of just before 1479 failed, even though the Siebenlist-Kerner chronology contained timbers from the same building. It was not until a temporary version of the 700-year chronology was available that the dating of the Milly samples was discovered. The end year of the chronology was 1596. The outermost sample had one year of sapwood allowing an estimated felling date in the range of 1613-1647. This timber was clearly not one of the original 1479 construction timbers. The 161-year series from the Milly timber formed a useful building block in the final north of France chronology. The 1596 end date is confirmed by direct comparison with the north of Germany (t = 4.4).

THE NORTHERN FRANCE CHRONOLOGY

Using a number of site chronologies including those mentioned above, a single chronology was constructed covering the span 1274-1979. The relative positions of the site chronologies is shown in Figure 5. It cross-dates with the independent Siebenlist-Kerner chronology with a t-value of 7.7 and with the German chronology with a t-value of 6.6. The chronology is presented in the form of arbitrary units (approximately equal to 0.02mm) in Table 4. It is intended simply as a temporary working chronology to simplify the process of dating other buildings in Northern France. It is weak in places where few trees are represented and would certainly not be suitable at this stage for climatic reconstruction studies. It is for this reason that it has not been presented in the form of indices (Fritts, 1976). Now that the outline chronology exists it should be easier to find more sites to increase the number of samples and to extend the chronology back to at least 1000 A.D.

In the last year attempts have been made to use this chronology for dating in Brittany. As might be expected from the strong climatic gradient from Brittany to the Loire Valley, the chronology is not ideal for dating as far west as Brittany, although timbers from a number of buildings have dated well. Work is now under way to construct a separate Brittany chronology.

Acknowledgements

Thanks are due to those who permitted sampling in buildings that they owned or were responsible for and to the British Academy who provided the finance for two seasons of fieldwork in France.

TABLE 4

Chronology for Northern France

DATE	RING WIDTHS IN ARBITRARY UNITS	NUMBER OF SITE CHRONOLOGIES
1274 -	176 155 106 143 143 149 195 138 207 234	1 1 1 1 1 1 1 1 1 1
1284 -	163 208 181 65 75 122 139 100 113 96	1 1 1 1 1 1 1 1 1 1
1294 -	117 139 136 157 159 95 88 82 127 65	1 1 1 1 1 1 1 1 1 1
1304 -	108 97 96 118 103 133 119 95 126 91	1 1 1 1 1 1 1 1 1 1
1314 -	127 134 144 113 108 71 80 92 108 96	1 1 1 1 1 1 1 1 1 1
1324 -	89 78 61 67 149 181 158 147 129 158	1 1 1 1 2 2 2 2 2 2
1334 -	129 107 57 68 57 96 119 110 143 136	2 2 2 2 2 2 2 2 2 2
1344 -	117 102 127 114 127 143 147 175 97 72	2 2 2 2 2 2 2 3 3 3
1354 -	81 86 93 132 138 155 119 99 114 148	3 3 3 3 3 3 3 3 3 3
1364 -	125 132 176 199 156 129 102 91 130 130	3 3 2 2 2 2 2 2 2 2
1374 -	129 118 103 103 101 112 118 114 98 146	2 2 2 2 2 2 2 2 2 2
1384 -	97 127 165 129 135 101 85 101 93 83	2 2 2 2 2 2 2 2 2 2
1394 -	93 104 97 99 137 126 107 95 133 163	2 2 2 2 2 2 2 2 2 2
1404 -	160 142 135 110 81 67 123 130 104 96	2 2 2 2 2 2 2 2 2 2
1414 -	99 90 85 94 106 73 102 99 69 62	2 2 2 2 2 2 2 2 2 2
1424 -	84 87 82 121 117 97 88 94 103 107	2 2 2 2 2 2 2 2 2 2
1434 -	65 76 105 158 226 116 100 139 93 91	2 2 3 3 3 3 3 3 3 3
1444 -	100 88 93 93 72 109 84 90 64 76	3 3 3 3 3 3 3 3 3 3
1454 -	83 94 98 97 97 95 86 77 54 70	3 3 3 3 3 3 3 3 3 3
1464 -	58 77 75 65 81 69 64 58 75 80	3 3 3 3 3 3 3 3 3 3
1474 -	58 56 59 66 61 80 88 85 81 57	3 3 3 3 3 3 3 3 3 3
1484 -	73 94 74 86 95 74 58 65 60 66	3 3 3 3 3 3 3 3 3 3
1494 -	61 60 90 88 76 79 95 85 98 80	3 2 2 2 2 2 2 2 2 1
1504 -	85 108 103 86 120 123 94 77 81 61	1 1 1 1 1 1 1 1 1 1
1514 -	64 102 57 41 66 47 49 55 58 72	1 1 1 1 1 1 1 1 1 1
1524 -	58 64 57 71 72 78 47 103 97 117	1 1 1 1 1 1 1 2 2 2
1534 -	101 115 113 102 80 98 77 100 82 77	2 2 2 2 2 2 2 2 2 2
1544 -	105 118 91 82 93 86 67 103 75 80	2 2 2 2 2 2 2 2 2 2
1554 -	75 102 73 56 59 55 100 80 110 97	2 2 2 2 2 2 2 2 2 2
1564 -	84 92 68 55 64 79 100 114 110 120	2 2 2 2 2 2 2 2 2 2
1574 -	88 94 84 79 56 73 70 69 88 83	2 2 2 2 2 2 2 2 2 2
1584 -	88 122 109 95 92 142 104 95 93 94	2 2 2 2 2 2 2 2 2 2
1594 -	121 118 142 104 94 96 114 113 95 73	2 2 2 1 1 1 1 1 1 1
1604 -	97 89 122 139 98 84 116 115 115 160	1 1 1 1 1 1 1 1 1 2
1614 -	170 125 140 167 167 154 170 184 156 115	2 2 2 2 2 2 2 2 2 2
1624 -	97 129 91 123 111 101 136 119 104 129	2 2 2 2 2 3 3 3 3 3
1634 -	140 132 125 127 122 120 134 106 131 140	3 3 3 3 3 3 3 3 3 3
1644 -	113 122 98 102 112 105 96 99 110 157	3 3 3 3 3 3 3 3 3 3
1654 -	130 94 96 94 134 95 111 151 120 129	3 3 3 3 3 3 3 3 3 3
1664 -	127 109 93 92 101 86 63 110 79 144	3 3 3 3 3 3 3 3 3 3
1674 -	115 120 72 110 95 73 58 59 106 99	3 3 3 3 3 3 3 3 3 3
1684 -	73 75 85 85 90 99 97 99 72 82	3 3 3 3 3 3 3 3 3 4
1694 -	74 72 75 85 86 92 99 100 71 99	4 4 4 4 4 4 3 3 3 3
1704 -	118 88 68 75 77 83 81 96 124 146	3 3 3 3 3 3 3 3 3 3
1714 -	104 105 89 91 88 72 100 93 93 76	3 3 3 3 3 3 3 3 3 3
1724 -	100 99 101 110 94 91 107 79 113 107	3 3 3 3 3 3 3 3 4 4
1734 -	86 131 93 107 120 98 100 79 69 87	4 4 4 4 4 4 4 4 4 4

TABLE 4 CONTINUED...

DATE	RING WIDTHS IN ARBITRARY UNITS	NUMBER OF SITE CHRONOLOGIES

```
1744 -   65  71 103 111 101  92  93  86  80  68    4 4 4 4 4 4 4 4 4 4
1754 -   91  71 102  78  90  74  66  65  65  93    4 4 4 4 4 4 4 4 4 4
1764 -   71  96  98  69  90 108 131 113 100 104    4 4 4 4 4 3 3 3 3 3
1774 -  109  99  99 113  86  79  99  95  85  95    3 2 2 2 2 2 2 2 2 2
1784 -   84  69  81  91  96 103  76  90 101  90    2 2 2 2 2 2 2 2 2 2
1794 -  100 116 113 120 103 101  96  98  79  77    2 2 2 2 2 2 2 2 2 2
1804 -   72 100  94  85  77  91  86  96  95 101    2 2 2 2 2 2 2 2 2 2
1814 -   99  78  97 100  85  83 107 117  85 108    2 2 2 2 2 2 2 2 2 2
1824 -  105  90  86  89 106 132 137 133  96  86    2 2 2 2 2 2 2 2 2 2
1834 -   74  72  98 111 109 114 100 123 108 123    2 2 2 2 2 2 2 2 2 2
1844 -  107 137 107 109 121  99 115  90  96 130    2 2 2 2 2 2 2 2 2 2
1854 -   93 121 112  83  72 103 135 119 106  87    2 2 2 2 2 2 2 2 2 2
1864 -  104  95 106 126 108 113  61 104  97 103    2 2 2 2 2 2 2 2 2 2
1874 -   68  84 101 100 127  98  82  96  98 112    2 2 2 2 2 2 2 2 2 2
1884 -   89  96 111  90 104  94 102 101  82  81    2 2 2 2 2 2 2 2 2 2
1894 -   95 108  79 116  97  81  76  82  84  86    2 2 2 2 2 2 2 2 2 2
1904 -   89  80  70  93  92  90 112  93 105 109    2 2 2 2 2 2 2 2 2 2
1914 -  107  90 109 115  97  99  99  64  85  98    2 2 2 2 2 2 2 2 2 2
1924 -  109 108 116 123 111 103 107 140 128 100    2 2 2 2 2 2 1 1 1
1934 -   81 106 105 108  67 107 100  95  79  75    1 1 1 1 1 1 1 1 1 1
1944 -   59  81  86 115 109  62 101 134  92  95    1 1 1 1 1 1 1 1 1 1
1954 -   90  77  69  87 156  91  93  98 103 115    1 1 1 1 1 1 1 1 1 1
1964 -   98 105 135 132 136 118 120 121  70  89    1 1 1 1 1 1 1 1 1 1
1974 -   80  90  73  93 109 104                    1 1 1 1 1 1
```

REFERENCES

Baillie, M.G.L. 1977a. The Belfast oak chronology to AD 1001. Tree-Ring Bulletin, 37 pp. 1-12.

Baillie, M.G.L. 1977b. An oak chronology for South Central Scotland. Tree-Ring Bulletin, 37 pp. 33-44.

Baillie, M.G.L. and Pilcher, J.R. 1973. A simple cross dating program for tree-ring research. Tree-Ring Bulletin, 33 pp. 7-14.

Bartholin, T.S. 1975. Dendrochronology of oak in Southern Sweden. Tree-Ring Bulletin, 35 pp. 25-29.

Becker, B., and Delorme, A. 1978. Oak chronology for central Europe. The extension from Medieval to prehistoric times. British Archaeological Reports International Series, 51 pp. 59-64.

Eckstein, D. 1978. Regional tree-ring chronologies along parts of the North Sea coast. British Archaeological Reports International Series, 51 pp. 117-124.

Fritts, H. 1976. Tree-Rings and Climate Academic Press.

Hollstein, E. 1980. Mitteleuropaische Eichenchronologie. Philipp van Zabern, Mainz am Rhein.

Hughes, M.K., Milsom, S.J. and Leggett, P.A. 1981. Sapwood estimates in the interpretation of tree-ring dates. Journal of Archaeological Science, 8 pp. 381-390.

Leboutet, L. and Colardelle, M. 1983. L'étude dendrochronologique de l'habitat médiéval immergé de Colletiére à Charavines (Isére): interprétations archéologiques. Archéologie Médiéval 13 pp. 131-154.

Munaut, A.V. 1978. La dendrochronologie une synthése de ses methods et applications. Lejeunia,. 91 pp. 1-47.

Pilcher, J.R., Baillie, M.G.L., Schmidt, B. and Becker, B. 1984. A 7272-year European tree-ring chronology. Nature 312, pp. 150-152.

Serre, F. 1978. Resultata dendroclimatiques pour les Alpes meridionales françaises. In: Evolution des Atmosphéres Planetaires et Climatologie de la Terre, ed. C.N.E.S. pp. 381-385, C.N.E.S. Toulouse.

Serre, F., Luck, H.B. and Pons, A. 1966. Premiéres recherches sur les relations entre les variations des anneaux ligneus chez Pinus halepensis Mill. et les variations annuelles du climat. Oecologia Plantarum, 1 pp. 117-135.

Trenard, Y. and Duchateau, J-L. 1985. Dendrochronologie du chêe dans la region de Paris. Dendrochronologia 3, pp. 9-23.

Problems of Dating and Interpreting Results from Archaeological Timbers

J. Hillam
Sheffield Dentrochronology Laboratory
Department of Archaeology and Prehistory
University of Sheffield
Sheffield S10 2TN

ABSTRACT

Although dendrochronology is now commonly used and widely accepted as a dating technique, and many tree-ring chronologies exist for numerous areas of Europe, there are still major problems involved with the method. This paper discusses, in general terms, some of the main problems encountered by dendrochronologists looking at archaeological timbers where there is little freedom of choice at the sampling stage.

Examples are given which illustrate these difficulties and show the possible limitations of dendrochronology as a tool for dating. In spite of the problems, many of the examples, such as the timbers from Coppergate in York or the Roman waterfronts in London, also demonstrate the success of the method and its potential in the field of archaeology.

INTRODUCTION

Although tree-ring analysis has been used as an archaeological dating method for over two decades in the British Isles, and many regional tree-ring chronologies are available, the success of the method still cannot be guaranteed. This paper discusses some of the limitations of the method and illustrates some of the problems, and how they have been overcome in practice at the laboratory at Sheffield. The two major recurring difficulties are those caused by lack of cross-dating and the estimation of felling dates in the absence of some or all of the sapwood rings. (The latter is discussed at length elsewhere in this volume, see Hillam et al this volume).

The Sheffield Dendrochronology Laboratory was established in 1975 primarily as a service laboratory for archaeologists, since when many hundreds of samples have been examined. Absolute dating has so far been confined to oak (Quercus spp), but other species have been used for relative dating (e.g. Morgan 1984). Size and age data have also been collected to provide information on woodland management and ecology (Morgan 1982). The percentage of timbers dated can be low or even zero for some sites since samples are collected with archaeology, not dendrochronology, in mind. As a result, much of the timber passing through the laboratory is totally unsuitable for cross-dating leaving us to extract what information we can about woodland ecology and management. Occasionally however long ring sequences are collected which allow the construction of chronologies of 200 to 500 years. In these cases the chronology building is regarded as a bonus. The number of timbers dated at Sheffield therefore cannot be compared with other laboratories simply because, in view of their archaeological importance we attempt to date samples which would normally be rejected (Hillam et al this volume).

DATING PROBLEMS: SOME EXAMPLES FROM THE U.K.

Long chronologies extending back to prehistoric times exist for Ireland and Germany (Pilcher et al 1984) whilst English chronologies go back to AD404, and also, through links with Ireland and Germany, cover the period 252BC to AD294. Baillie (1983a) has suggested that the framework of existing chronologies is sufficient to date any chronology from the British Isles, at least within the historic period, providing site masters can be obtained. However, this is not always the case.

Apart from the problem of matching single timbers to form a floating chronology which may then be dated absolutely by cross-matching with a reference chronology, other difficulties include:

1: The absence of full sapwood, which may preclude the precise dating of the structure.

2: The use of imported timber which may not cross-date with native timbers from the same site.

3: The re-use of timber which may give a spuriously early date for a structure or context.

Some of these problems are illustrated by the case histories which follow.

1: Hartlepool and Hamwic

'Ideal' conditions for archaeological dating are illustrated by the case of Hartlepool and Hamwic. The timbers provided long ring sequences which subsequently matched to form site masters, which were satisfactorily dated against reference chronologies.

At Hartlepool, six timbers were examined from the lining of a dock excavated by Blaise Vyner in 1981. The construction date was unknown although the dock appeared to go out of use in the late 13th/early 14th century. The samples had 106 to 256 rings, and four cross-matched to give a site chronology of 362 years. This sequence dated to AD851-1212 by comparison with several British chronologies. In addition, two timbers retained their full sapwood, including the last season's latewood which indicated felling in the winter of AD1212/13. The dock was probably constructed after this since seasoning would not be necessary, nor was it common practice unless the timber was intended for panels or furniture.

In the case of Hamwic the only difficulty arose from the absence of full sapwood, which precluded the precise dating obtained at Hartlepool. The timbers from a Saxon well in Southampton were excavated by Phil Andrews. This particular well, structurally unique for Hamwic, was 4m deep so that the lowest timbers were preserved. The site master dated to AD458-710 but no timbers had retained all their sapwood which meant that felling dates had to be estimated. All but one timber was estimated as AD709$\pm$9: the remaining timber, which may have been added later was felled some time after AD 733.

Table 1: Dating the Coppergate 'warehouse'.

reference chronology	t-values at position:	
	AD 914-1011	AD 1138-1235

a) Results obtained in 1980

Dublin (Baillie 1977)	4.5	3.6
Exeter (Hillam 1980)	3.8	3.7
Germany, Schleswig (Eckstein pers comm)	3.1	2.1
Germany, Trier (Hollstein 1980)	2.1	3.7

b) Results of 1983 which confirmed the 914-1011 date

Lincoln (Laxton et al 1982)	7.5	no overlap
England (Baillie & Pilcher pers comm)	5.2	2.8
Britain (")	5.4	2.8

Figure 1. Waterlogged timber from Ipswich showing included sapwood. (Photo: PW Kingsland)

2. Coppergate

Unlike Hartlepool and Hamwic, complex urban sites such as Coppergate are much more difficult and time-consuming. Over a seven year period this site produced more than 400 tree-ring samples, of both mediaeval and viking origin. Dating the material was hampered by two types of problem, one connected with the provision of samples, and the other related to the samples themselves.

As regards sample provision three major problems arose:

1: Information about timber context and phasing was often lacking, due to the complexity of the site and the long period of excavation. Consequently timbers could not be grouped according to structure.

2: Some of the better timbers were not sampled until very recently following decisions about which timbers to conserve.

3: Some timbers were rejected from analysis early on, only to require reanalysis later as site chronologies were slowly prepared.

Problems connected specifically with the samples included:

1: The rejection of about 120 samples with too few or too narrow rings (at first samples with less than 50 rings were rejected: later, samples with more than 30 rings were measured provided they included sapwood).

2: Although four of the longest sequences matched to give a master of over 300 years this failed to match with either dated reference curves or other Coppergate sequences, including the ring sequences from the same structure.

3: Some of the ring patterns were not unique and matched in more than one position, which meant that independent chronologies had to be generated in order to select the correct match. The clearest example of this comes from seven timbers from the 'warehouse' which crossmatched to give a 98 year master. In 1980 this seemed to cross-date in two positions and the correct position could not be determined from either visual or statistical inspection. In spite of archaeological grounds casting doubt on the later date, tree-ring matches MUST be accepted on the positive evidence of cross-dating, rather than the negative evidence of archaeological unacceptability. The former came in 1983 with the provision of the Lincoln Cathedral chronology (Laxton et al 1982) which cross-dated well over the period AD914 -1011. Subsequently other reference curves have confirmed this date, whilst the Lincoln chronology has assisted with other sequences from Coppergate (Table 1).

Another problem highlighted by the Coppergate timbers was that of foreign trade. Dendrochronologists have tended towards the naive assumption of local origins for their timber, possibly a single woodland near to the structure for which it was used. In fact the timber trade was probably well developed by Roman times and the great variety of the Anglo-Scandinavian timbers from Coppergate, as shown by the size and age of the trees and the width and pattern of their growth rings, indicates that the wood used in a single structure could come from several sources. The timber trade was even better developed in the middle ages (e.g. Rackham

Figure 2. Building timber from South Yorkshire showing the variation in the number of sapwood rings. Note that the heartwood reaches almost to the bark edge on the left. (Photo: PW Kingsland)

1980), as the tree ring analysis of timbers from places such as Bristol, Ipswich and Norwich is beginning to show (Hillam 1985). Timber imports were large scale (Baillie et al 1985) and ports like Bristol both imported and distributed a variety of timbers. This causes problems for tree-ring dating because timbers from the same structure, if from different source areas, may not cross-date, and there is no biological basis for constructing a single master curve. This problem occurred in the dating of the Courts site, Norwich, excavated by Brian Ayers in 1981. From 50 timbers only ten usable ring sequences were produced of which only one could be dated.

Thus Coppergate provided many problems which were ultimately overcome (see Hillam 1985 for a summary of some of the results). In addition it demonstrated the advantage of dating several timbers from a single structure rather than single timbers, a conclusion noted by Eckstein in his work on timbers from Hedeby and Schleswig (e.g. Eckstein and Wrobel 1982).

INTERPRETATION PROBLEMS

The other major problem concerns the interpretation of tree-ring dates. These refer to the date of a timber's outer ring which is not necessarily the same as the felling date of the timber, less still the construction date. When the timber is complete to the bark edge then the felling date is obviously known. But more usually some wood, particularly the sapwood, which is the outer part of the oak tree, and distinguishable from the heartwood by a colour change and the absence of tyloses, was trimmed off when the timber was converted into a plank or beam. Felling dates must then be estimated, and the accuracy of the estimates depends on the presence or absence of sapwood, and on how precisely the number of missing sapwood rings can be estimated. The number of sapwood rings is regarded as relatively constant. However, there are five major problems:

1: Included sapwood may occur (Figure 1) which could lead to a 'too early' estimate of the felling date.

2: Waterlogged wood may be discoloured at the edges and the vessels may lack, or seem to lack, tyloses. This may occasionally be mistaken for true sapwood.

3: Waterlogged wood often shows a lighter band of rings just before the true sapwood is reached. If the latter is present there is no confusion, but when absent the lighter rings may be misidentified as sapwood. The presence or absence of sapwood is therefore not always certain, and in cases of doubt it is usually best to assume its absence.

4: The number of sapwood rings in oak is more variable than once thought, varying both within and between trees. There are more sapwood rings towards the crown of the tree (though position in the tree is an unknown factor in archaeological timbers) (Hughes et al 1981), and the number of sapwood rings varies around the circumference (Figure 2). This variation is usually small, but may be up to 20 rings (Figure 3) and the heartwood may reach the bark (Baillie 1982 plate 1e). Although a localised occurrence, if this occurred on a thin board then the estimated felling date would be much later than the true one.

5: Finally there is increasing evidence that sapwood numbers increase east to west across north-west Europe (Hillam et al, this volume). The

FISKERTON

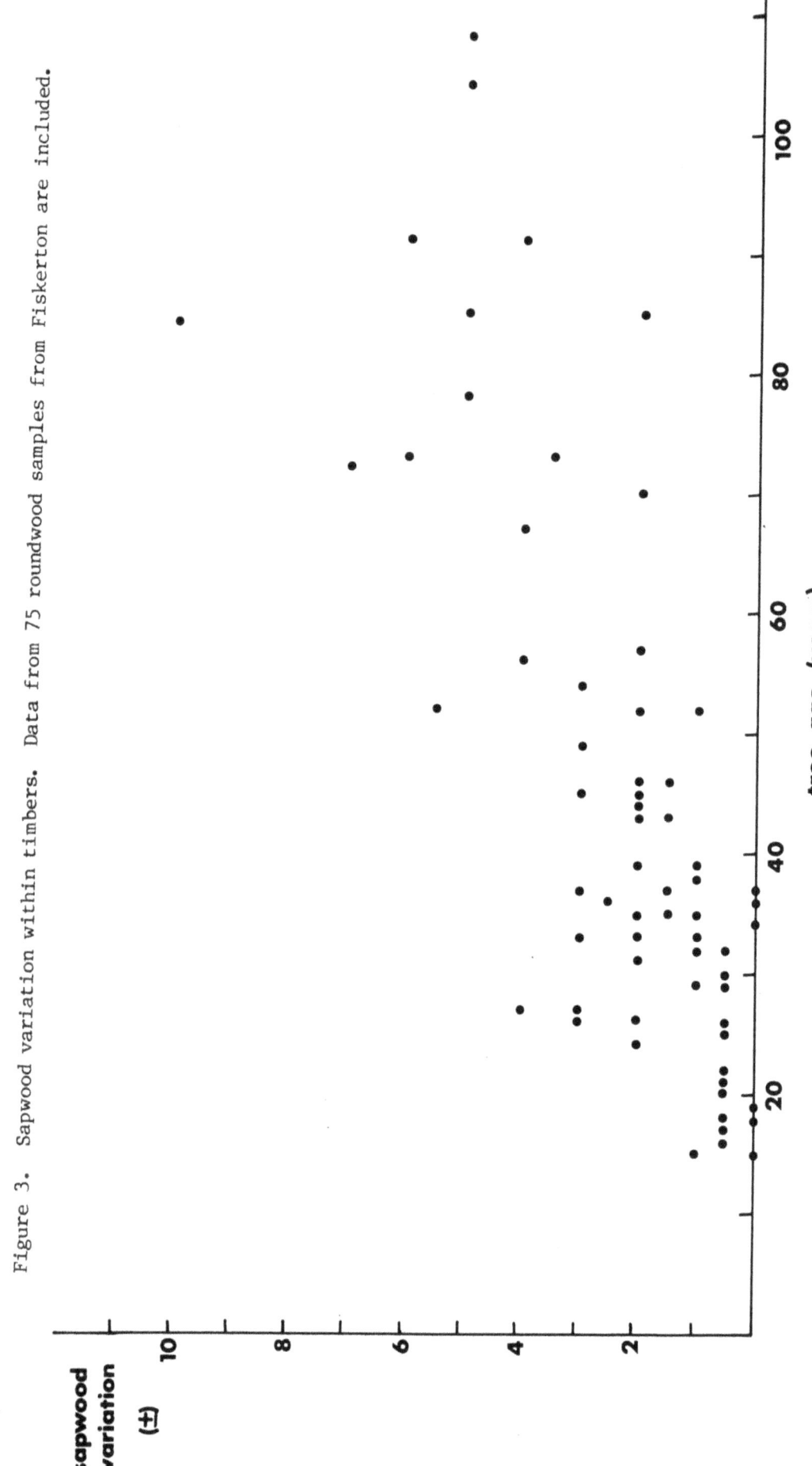

Figure 3. Sapwood variation within timbers. Data from 75 roundwood samples from Fiskerton are included.

implication of this for structures built of imported timbers from the Baltic, for example, is that felling date estimates are likely to be too late, i.e. too near the present. (Baillie et al) 1985).

The implications of these problems for archaeology seem fairly serious, especially where a site is complex and tree-ring results cannot distinguish between the various phases of building because the time span between them is less than the likely variation in sapwood number. This can be seen clearly with some of the Roman timbers from the Thames waterfront in the City of London (Bateman and Milne 1983). The three excavations around the site of Roman London Bridge (Pudding Lane and Peninsular House to the east, and Miles Lane to the west of the bridge) have produced the remains of several quays and warehouses, which have provided several hundred timbers for analysis. Very few of the timbers had sapwood (Figures 4, 5) and many had an unknown amount of heartwood rings missing. The tree-ring results alone therefore could not distinguish between the various phases, unlike the results from Coppergate (for example, compare Figure 6 with Figure 7), where the felling dates were eventually estimated with some precision even though the bark edge was rarely present.

The London results, however, provide a firm foundation for the Roman waterfront chronology, even though they do not give precise construction dates. They demonstrate that none of the waterfront structures at Pudding Lane or Peninsular House could have been built before AD60, but that they were all in place by about AD100. In addition, there were only circa 20 years between the construction of quays 1 and 2, whilst the bridge pier was built shortly after quay 1. The latter results are deduced from the tree-ring dates and the stratigraphy, again demonstrating the need for consultation between dendrochronologist and archaeologist.

The Roman waterfront results also show another valuable aspect of dendrochronology. It enables structures from different sites to be linked together in a relative, and sometimes absolute, framework. Although only four of the dated timbers from Miles Lane had sapwood, it was possible to show that its waterfront complex was built before the equivalent structures to the east of the Roman bridge (Thus the Miles Lane quay and revetment are similar in date to quay 1, but earlier than quay 2 at Pudding Lane).

CONCLUSION

The problem of interpreting tree-ring results - at least to the satisfaction of the excavator who would like precise felling dates, is one that cannot be solved. Oak sapwood is undoubtedly variable, both within and between trees, and it was often removed because of its susceptibility to insect or fungal attack. Estimation of felling dates in the absence of some, or all of the sapwood will therefore always be approximate, especially where the site is complex with many phases, and where timber is re-used or structures repaired.

The Hartlepool results show that tree-ring dating can provide felling dates that are exact to the year. If bark edge is not detectable, but there are sufficient samples with some sapwood, the felling dates can be estimated with some precision, as at Hamwic and Coppergate. Where there are few timbers with sapwood, such as the Roman waterfront sites in London, tree-ring analysis can still provide a firm dating framework, which can be refined by additional evidence, such as that provided by stratigraphy.

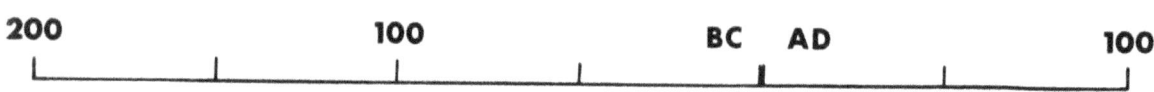

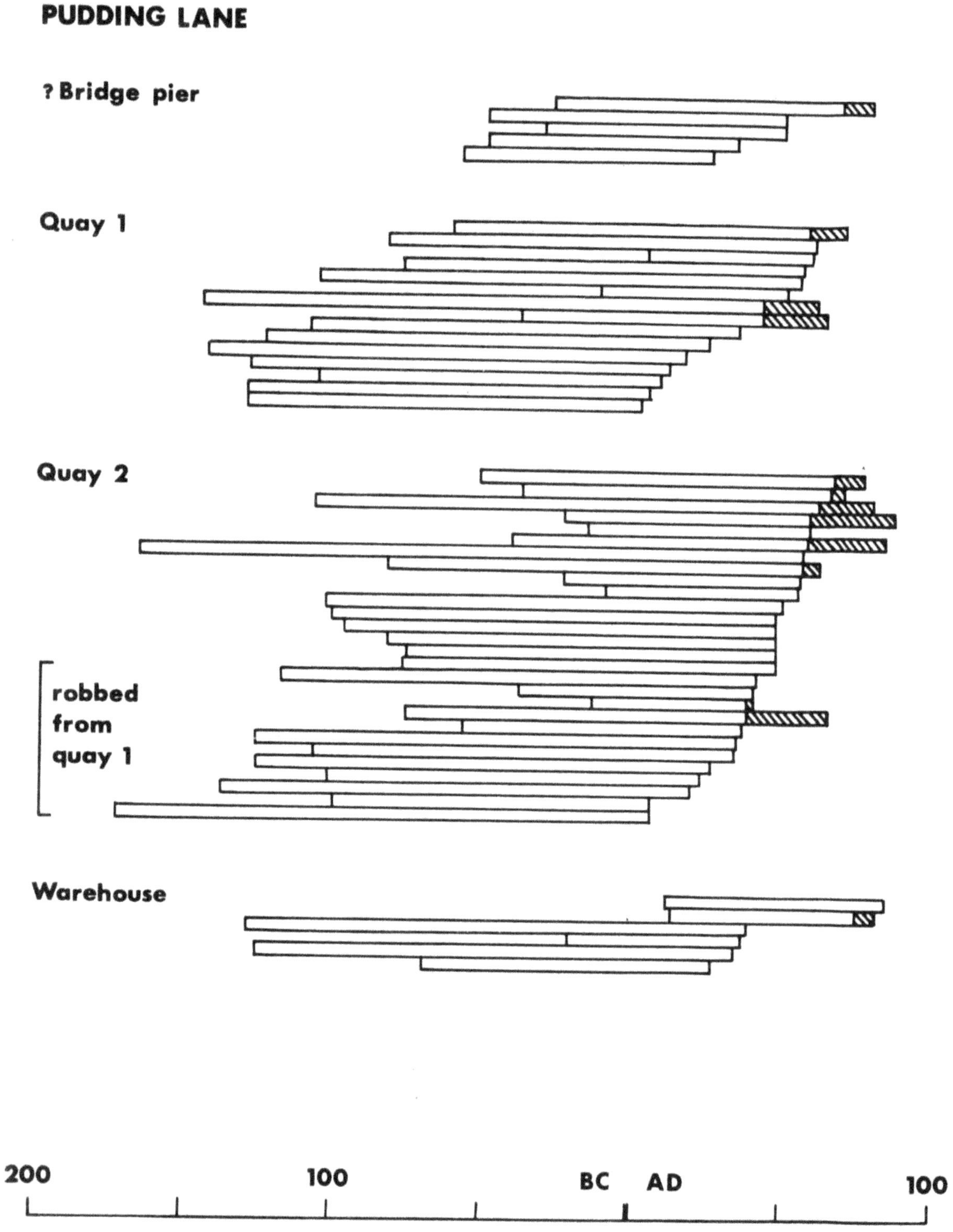

Figure 4. Bar diagram showing relative positions of ring sequences from two of the Roman waterfront sites in London. Open bar = heartwood; hatching = sapwood.

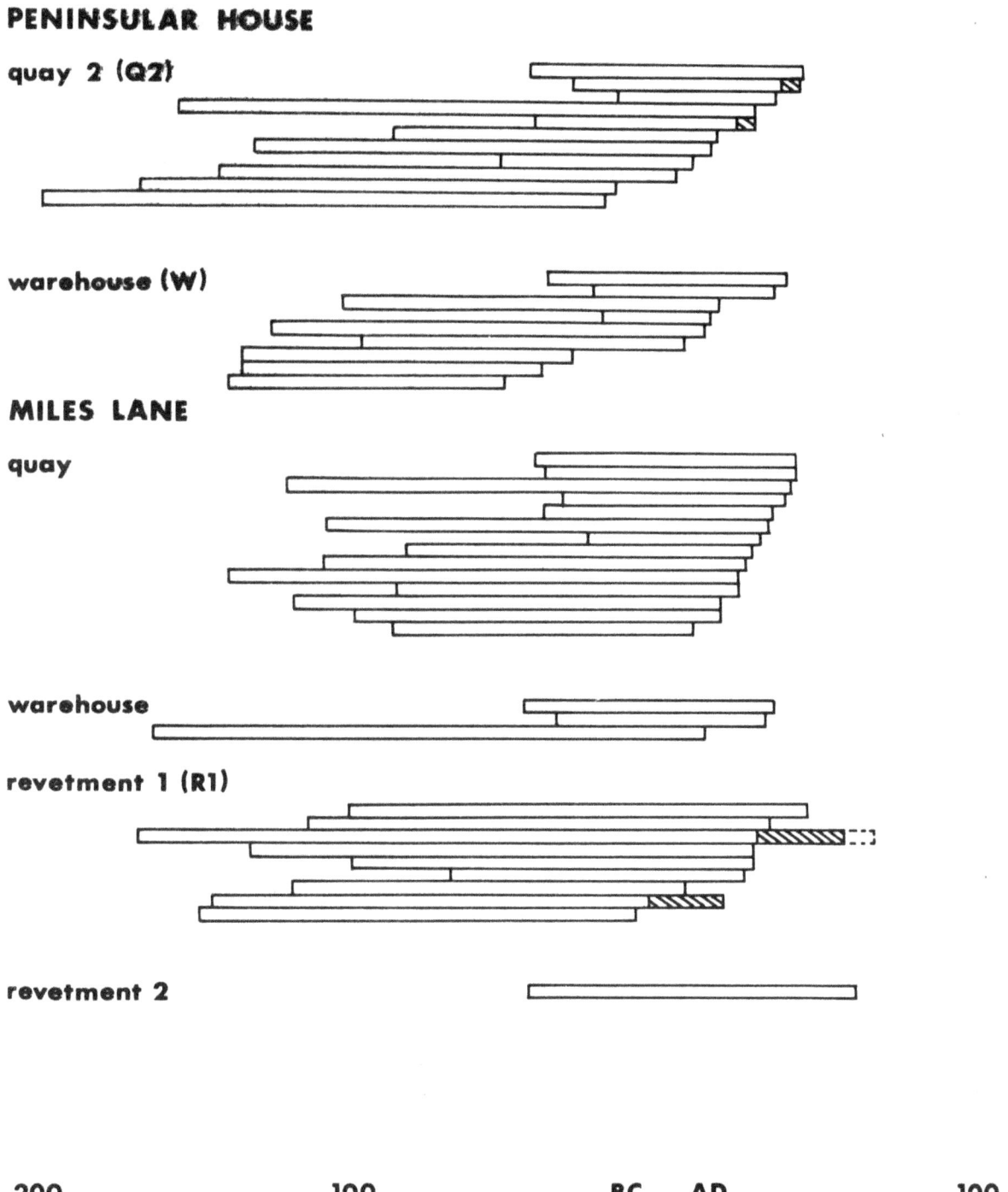

Figure 5. Bar diagram showing relative positions of some of the Pudding Lane ring sequences. Pudding Lane and Peninsular House are to the east of Roman London bridge; Miles Lane is to the west.

The problem of interpretation is often alleviated by taking as many samples as possible: at Coppergate, where all the timbers were kept for possible conservation, more and more samples were cut until sufficient timbers with sapwood were found to date all the sunken buildings. The simplest policy is to sample all the timbers during excavation (although this was not known at the start of the Coppergate excavation). This increases the chances of finding timbers with complete sapwood. It also makes the estimation of felling dates more precise because interpretation is based on more samples. The excavator can help by providing the dendrochronologist with as much information as possible, and as soon as possible, to enable timbers to be grouped according to phases or structures. The results for the major sites at London and York would not be as exact without the additional information from the excavators.

These recommendations will also help to overcome the problems caused by lack of crossdating. If the dendrochronologist is involved at the start of an excavation, he or she can then advise on sampling as well as collecting information about the samples and their associated timbers. If as many samples as possible are taken, there is a better chance of producing a site master curve, which will be easier to date than individual ring sequences.

The consequences of the timber trade, and the complex nature of timber transport and timber re-use, are only just being realised by dendrochronologists (e.g. Baillie 1983b). As long as the problem is acknowledged, then it should not matter because, even if timbers are imported from the Continent, there are plenty of Continental reference chronologies by which to date them.

In the laboratory, techniques are continually being improved. Work continues on the statistical side of crossdating, and the number and quality of reference chronologies in Britain and Europe is increasing all the time. The use of short ring sequences is also being studied, and the results are encouraging (Hillam et al this volume). Provided sequences are well-replicated, there is no reason why timbers with as few as 30-50 rings cannot be dated. In this case, crossdating success can sometimes be improved if the rings along two radii are measured and then averaged.

Although the problems and limitations of dating archaeological timbers may at first seem daunting, many can be overcome by sensible sampling. It is rare for tree-ring work to produce no dates, and the interpretation of them can often be refined further by looking at additional evidence from an excavation. At its best, dendrochronology can provide a firm, and sometimes very precise dating framework, even for the most complex site.

Acknowledgements

The Sheffield Dendrochronology Laboratory is funded by the Historic Buildings and Monuments Commission for England. I am also grateful to all the archaeologists involved with the sites mentioned in the text, particulary to Richard Hall and Gustav Milne who have supplied much information about the Coppergate and London timbers respectively. I would also like to thank Ruth Morgan and Ian Tyers for their comments on the article.

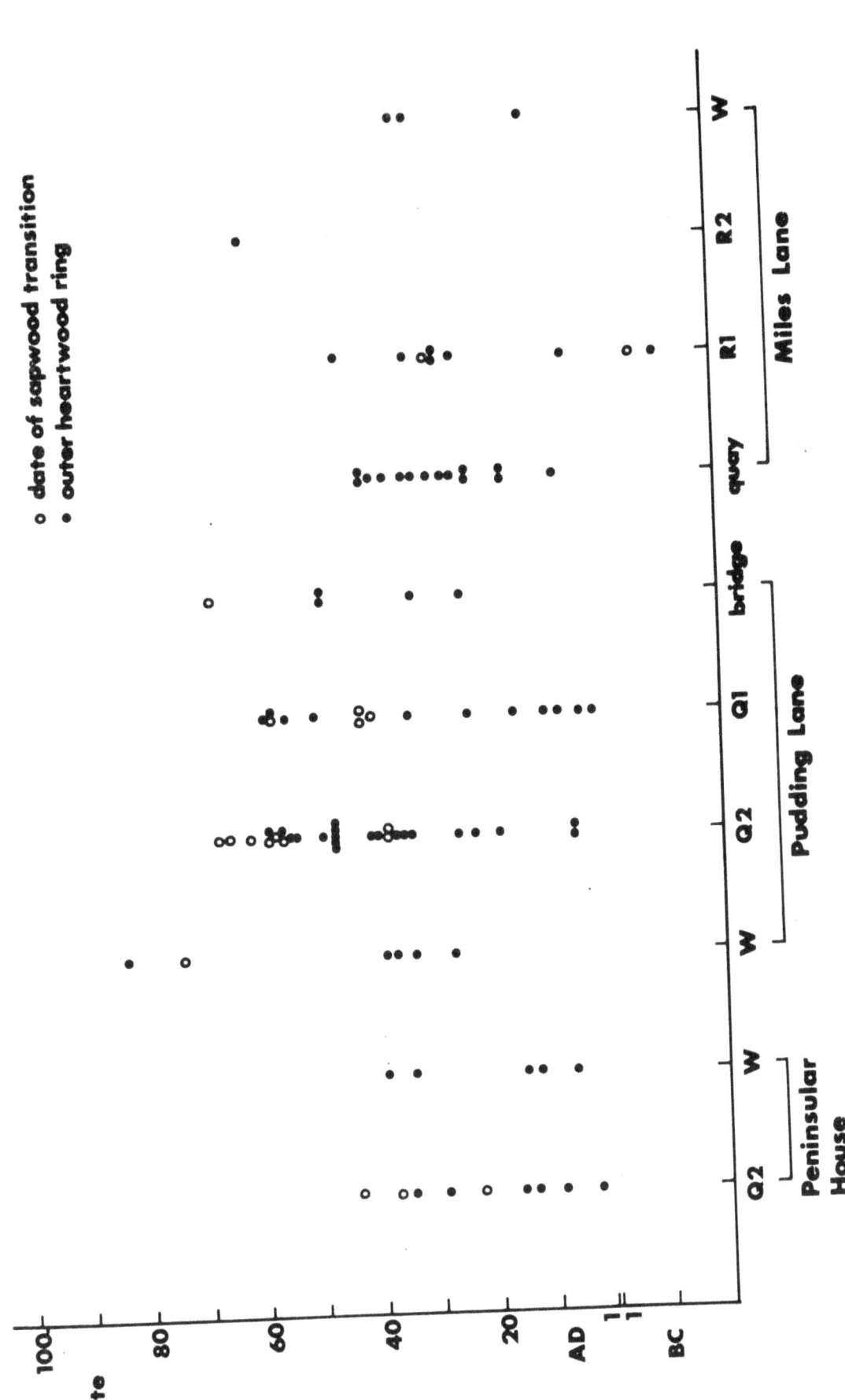

Figure 6. Dates of the outer heartwood rings or of the heartwood-sapwood boundary for the Roman waterfront structures from London. These could not be interpreted without additional stratigraphical evidence from the excavator. Compare these results with those from Coppergate (Figure 7).

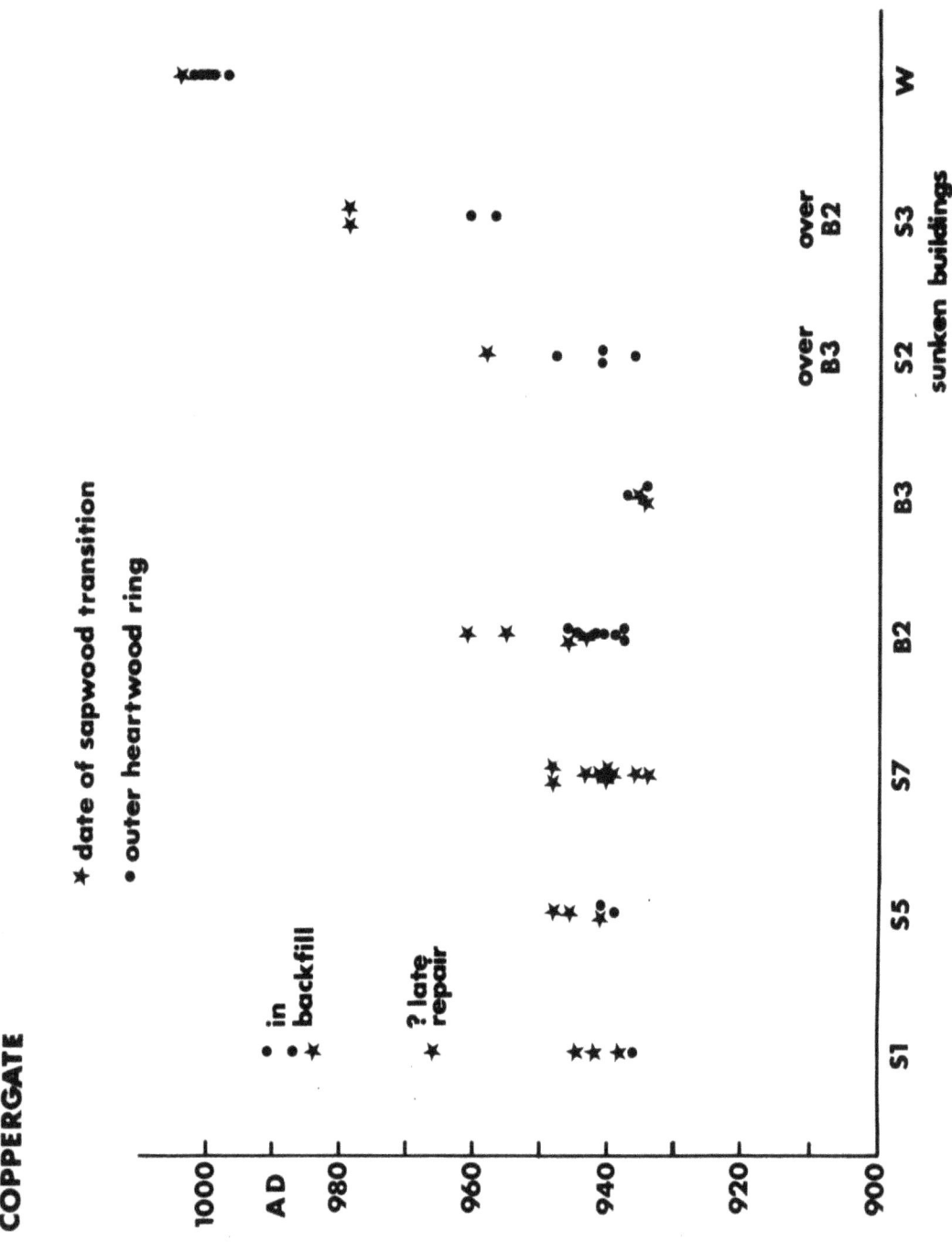

Figure 7. Dates of the outer heartwood rings or of the heartwood-sapwood boundary for each of the Coppergate sunken buildings. S = structure; B = building; W = warehouse. Evidence of repair to two workshops is apparent (S1, B2). Structure 2 was probably built about the same time as these repairs.

REFERENCES

Baillie, M.G.L. 1982. *Tree-Ring Dating and Archaeology*. London: Croom Helm.

Baillie, M.G.L. 1983a. Is there a single British Isles oak tree-ring signal? In: *Proc 22nd symposium on Archaeometry* (eds Aspinall, A. & Warren, S.E.), Bradford, pp. 73-82.

Baillie, M.G.L. 1983b. Development of Tree-Ring Chronologies. In: *Dendrochronology and Archaeology in Europe* (eds Eckstein, D., Wrobel, S. & Aniol, R.W.), Hamburg: Max Wiedebusch, pp. 33-48.

Baillie, M.G.L., Hillam, J., Briffa, K.R. & Brown, D.M. 1985. Re-Dating the English Art-Historical Tree-Ring Chronologies. *Nature* 315, pp. 317-319.

Bateman, N. & Milne, G. 1983. A Roman Harbour in London: Excavations and observations near Pudding Lane, City of London 1979-82. *Britannia* 14, pp. 207-226.

Eckstein, D. & Wrobel, S. 1982. Dendrochronology in Europe - with special reference to Northern Germany and Southern Scandinavia. *PACT* 7.1, pp. 11-25.

Hillam, J. 1985. Recent tree-ring work in Sheffield. *Current Archaeology* 9(1), pp. 21-26.

Hillam, J., Morgan, R.A. and Tyers, I. 1987. (this volume) Sapwood Estimates and the Dating of Short Ring Seqeunces. pp. 165-185.

Hughes, M.K., Milsom, S.J. & Leggett, P.A. 1981. Sapwood estimates in the interpretation of tree-ring dates. *Journal of Archaeological Science* 8, pp. 381-390.

Laxton, R.R., Litton, C.D., Simpson, W.G. & Whitley, P.J. 1981. Tree-ring dates from some East Midland buildings. *Transactions of the Thoroton Society* 86, pp. 73-78.

Morgan, R.A. 1982. Current tree-ring research in the Somerset Levels. In: *Archaeological Aspects of Woodland Ecology* (ed Bell, M. & Limbrey, S.) Oxford: British Archaeological Reports S146, pp. 261-277.

Morgan, R.A. 1984. Tree-ring studies in the Somerset Levels: the Sweet Track 1979-1982. *Somerset Levels Papers* 10, pp. 46-64.

Pilcher, J.R., Baillie, M.G.L., Schmidt, B. & Becker, B. 1984. A 7,272-year tree-ring chronology for western Europe. *Nature* 312, pp. 150-152.

Rackham, O. 1980. *Ancient Woodlands. Its History, Vegetation and Use in England.* London: Edward Arnold.

The Belfast CROS Program - Some Observations

J.R. Pilcher and M.G.L. Baillie

Palaeoecology Centre
Queen's University
Belfast

ABSTRACT

Empirical tests are presented that highlight the dangers of attempting to cross-date short samples. Computer cross-dating of short samples frequently produces either no results or erroneous results. The length of sequence needed for reliable dating increases as the distance between sites increases and the number of trees forming each chronology decreases.

INTRODUCTION

At its most fundamental, dendrochronology is about the correct relative placement of two sections of tree-ring pattern. If we were to take the ring pattern of a living tree in a forest we should be able to date the ring pattern of a stump from the forest against the living tree pattern, provided that the species is the same. Imagine we perform such an exercise and we call our living tree A. The dendrochronologist is attempting to date the pattern for stump B by comparing it visually with the pattern from tree A and assessing every possible position of overlap for similarity. Let us imagine in this exercise that A ends in 1985 and the dendrochronologist finds an acceptable match with the last ring of B at 1916.

```
|_____|   A (living tree)


|_____|           B (stump of tree)
                                       |    |
                                     1916  1985
```

This dating seems reasonable, but as with all human judgements it tends to be somewhat subjective. How might the dendrochronologist verify his '1916 dating'? Presumably if another observer could arrive at the same relative position for B independently this would reinforce the idea that there was something meaningful about the 1916 position. A more efficient approach is to make use of a computer program as an independent test of similarity. The use of the computer confers the advantage that the same test can be applied at every overlap position and it also ensures that every possible position is being checked.

The two commonly used computer programs for assisting with cross-dating are the Hamburg Gl% (Gleichlaeufigkeits) or 'w' program (Eckstein and Bauch,

1969) which calculates the percentage of years in which the two tree-ring curves show the same trend, and the Belfast CROS program (Baillie and Pilcher, 1973) which calculates 't' values. Both calculate statistics of similarity which attempt, albeit crudely, to model what the dendrochronologist is doing when he looks for similarity between two ring patterns. Let us state clearly that neither Gl% nor 't' is ideal. Most criticism of CROS has centred, not on its ability to pick out correct matches, but on the derivation of true significance levels associated with the t-values (Barefoot et al, 1978; Orton, 1983). Other workers have derived improved, but more complex, methods of computer cross-dating (Laxton and Litton, 1982, 1983; Munro, 1984), however the original programs have the proven advantage that it is easy to understand what the computer is doing and both are capable of specifying the best matching position correctly between two contemporary ring patterns.

There is however one important point about any computer matching procedure. In order to be workable in practice some form of mutual veto has to exist between the human observer and the computer. Only when the human observer likes the look of a match and the computer rates that same position, not just as a good match, but as the best match out of all possible overlap positions do we have an ideal case.

There is a further qualification with respect to the computer's "best correlation position". What happens if the computer picks out the correct matching position with a 't' value of 4.6 (overlap 195 years) but a higher correlation, say t = 5.3 occurs at a short overlap position (22 years). In this case the correct match is not giving the highest absolute value but there is little doubt that the 4.6 value is more significant. When the program was devised we set an arbitrary starting overlap between the ring patterns of 14 years rather than the 60-80 years we would now choose on the basis of empirical tests such as those below.

The published version of the Belfast CROS program (Baillie and Pilcher, 1973) takes two sections of ring pattern, A and B (either ring patterns of individual trees or averages of several individuals i.e. master chronologies), removes trends and then normalises the data. This leaves the high frequency information - the year to year detail - which is the most important matching component in dendrochronology. The program then sets an overlap position and calculates a correlation coefficient, r, which is converted to 't' in order to introduce some measure of overlap length. 't' is calculated in this way for every overlap position. At mis-match positions the time series should be unrelated and the correlation should approximate to zero. In practice most of the mis-match positions give 't' values between -3 and +3. At a match position, i.e. where the two ring patterns grew over the same period of time, the correlation coefficient should be high and 't' should be greater than 3.5. To be really useful a true match position should have a 't'-value significantly greater than the next highest positive value. To illustrate this, the 't'-values for all possible positions of overlap of two ring series were calculated and are presented as a histogram in Figure 1 (see also Orton, 1983 Figure 2). As can be seen they approximate to a normal distribution with a mean of zero and with 99% of the values falling between -3 and +3. The position selected by eye gives a t-value of 7.6 (for 322 yrs overlap) which stands out clearly from the normal distribution. Such a t-value indicates that the 1916 dating selected by the dendrochronologist is also selected by the computer as the best correlation between the two ring patterns. In a blind dating exercise where the ring patterns are compared by the computer before being

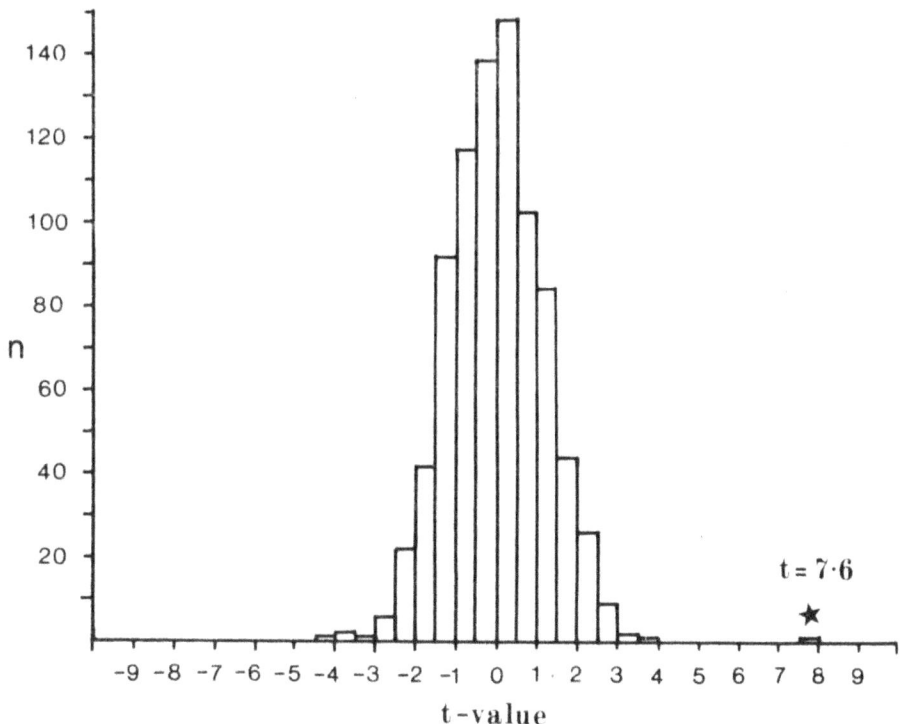

Figure 1. Histogram of t-values calculated as in Baillie and Pilcher (1973), for all positions of overlap between the ring pattern of a stump from Sherwood Forest and a living tree. The correct cross-dating position established visually is backed up by the uniquely high t-value of 7.6.

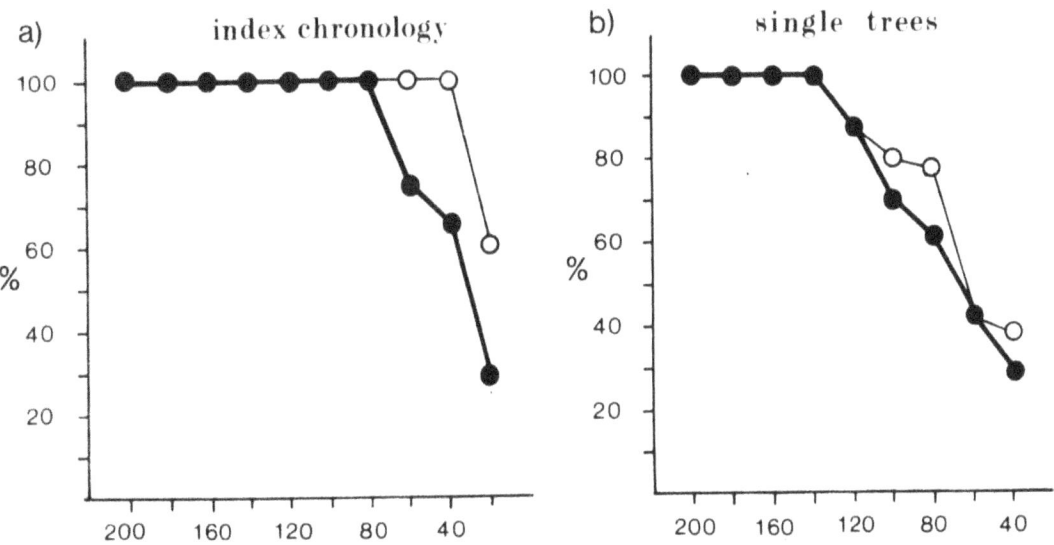

Figure 2. Percentage success rate graphs for cross-dating blocks of variable size. Solid circles are for UNIQUE dating (t greater than 3.0 and highest value), open circles are for INDICATED dating (t-value greater than 3.0 but not the highest t-value).
    (a) Comparison of index chronologies from Shanes Castle and Castlecoole, distance 96 km.
    (b) Comparison of individual trees from Castlecoole with index chronology from Shanes Castle.

checked by the dendrochronologist, this t-value would indicate a relative position <u>well worth investigating</u> by the human observer. No particular value of 't' automatically says that the true matching position has been found. On the basis of vast amounts of empirical data, values of 't' greater than 6.0 are seldom in error. However even with such high values it will always be necessary to check the overlaps visually as otherwise the dendrochronologist is abdicating his responsibility to the machine!

In the above example we find that there is only one correlation position with 't' greater than 3.5 and that is for B ending in 1916 with 't' = 7.6. This is however a near ideal case with both trees growing in the same forest. We term such dating UNIQUE where CROS clearly indicates the true position. A less satisfactory situation is where CROS gives a number of correlation values greater than 3.0 including the true position which may subsequently be picked out by the human observer. This is not unique dating but clearly CROS has reduced the possibilities and the true position is INDICATED as one of a limited number of possibilities.

EMPIRICAL TESTING OF CROS IN DIFFERENT SITUATIONS

The following empirical tests demonstrate the effects of chronology length and distance between sites on the performance of CROS. Long chronologies and long ring patterns from individual trees were divided into subsections. The precise date of each subsection was known. Each subsection of the test series was run through CROS against the full length of the other series. In this way we can observe where CROS breaks down. The results were scored as:

(1) UNIQUE dating (i.e. CROS gives highest t-value at correct date)

(2) INDICATED dating (t-value greater than 3.0 at correct date but not highest t-value).

The first two tests use ring patterns from two recently constructed living-tree chronologies from Shanes Castle, County Antrim and Castlecoole, County Fermanagh in Northern Ireland. The sites are 96 km apart in a NE-SW direction. We assigned Shanes Castle as the absolute chronology and tested subsections of the Castlecoole chronology. Both chronologies were index chronologies combining the ring patterns of many trees. Figure 2a shows the effect of chronology length on "dating success". With these high quality, well replicated chronologies unique dating is given by CROS down to chronology lengths of 80 years. However when a similar test is carried out using <u>individual trees</u> from Castlecoole against the Shanes Castle chronology the picture is quite different (Figure 2b). Here unique dating is only given for subsections longer than 120 years. This is much more akin to the normal archaeological dating situation where individual samples are tested against an existing chronology from nearby. One should not expect unique dating if the sample is shorter than 120 years. In fact the correct position is not even indicated in half the samples of less than 80 years.

Long distance cross-dating

In the following tests well replicated index chronologies were used, but the distances were greater. Figure 3a shows the effect of decreasing block

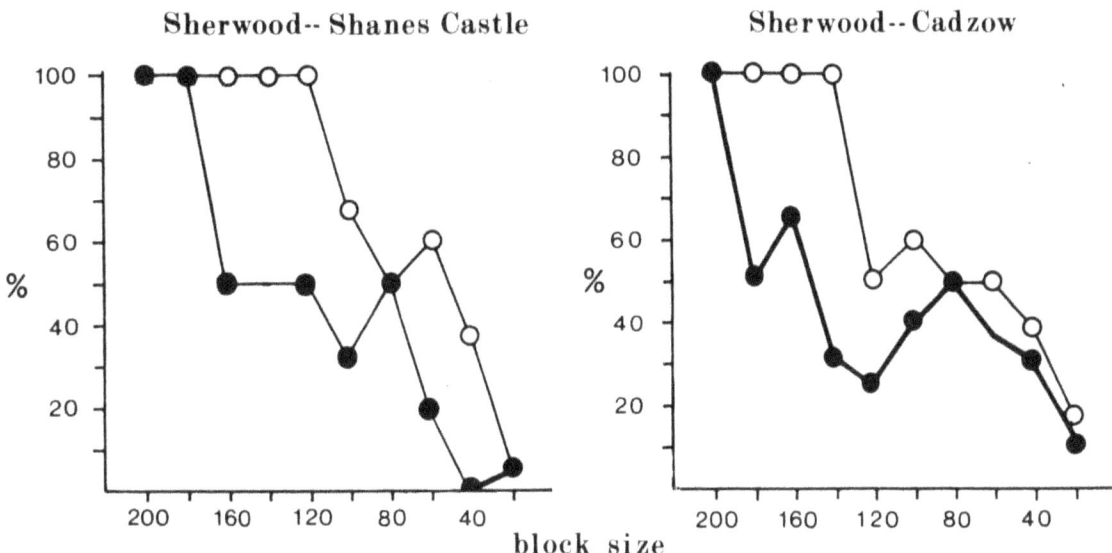

Figure 3. Percentage success rate graphs for varying block size as in Figure 2.
    (a) Index chronologies from Sherwood Forest and Shanes Castle, distance 400 km.
    (b) Index chronologies from Sherwood Forest and Cadzow, distance 260 km.

size on comparison of chronologies from Shanes Castle in Northern Ireland and Sherwood Forest. The sites are 400 km apart. The full length of chronologies cross-date well with a t-value of 5.8, but unique dating is only found down to a block size of 180 years. Below 120 years the correct position is not indicated for many blocks. The second example of this type is a similar comparison of Cadzow forest in Central Scotland with Sherwood Forest, a distance of 260 km (Figure 3b). Although the full chronology gave a t-value of 7.8, unique cross-dating breaks down below 200 years and even indicated dating was not found in 50% of blocks less than 140 years.

CONCLUSION

The message is simple. The results given above are for a few illustrative tests, which could have been extended to include multi-site chronologies and their performance against individual ring patterns and site masters at different distances. There seems to be little point and users of CROS must test it out for their own purposes and to their own satisfaction. The tests we have included suggest that you shouldn't expect CROS (or any other similar statistical test) to provide the correct date even under ideal conditions for site chronologies of less than 80 years and individual trees less than 120 years. Where the sites are hundreds of km apart this minimum length is likely to be as much as 200 years.

REFERENCES

Baillie, M.G.L. and Pilcher, J.R. 1973. A simple cross-dating program for tree-ring research. <u>Tree-Ring Bulletin</u> 33, pp. 7-14.

Barefoot, A.C.,, W.L. Hafley and J.F. Hughes 1978. Dendrochronology and the Winchester excavation. In <u>Dendrochronology in Europe</u> (B.A.R., International Series 61), ed. by J.M. Fletcher, pp. 162-171.

Eckstein, D. and Bauch, J. 1969. Beitrag zu rationalisierung eines dendrochronologishen Verfahrens und zu Analyse siener Aussagesicherheit. <u>Forstwis Centralbl</u>. 88, pp. 230-250.

Laxton, R.R. and Litton, C.D. 1982. Information theory and dendrochronology. <u>Science and Archaeology</u> 24, pp. 9-24.

Laxton, R.R. and Litton, C.D. 1983. Information theory and dendrochronology: (II) The effect of pre-whitening. <u>Computer Applications in Archaeoology</u>.

Munro, M.A.R. 1984. An improved algorithm for crossdating tree-ring series. <u>Tree-Ring Bulletin</u> 44, pp. 17-27.

Orton, C.R. 1983. The use of Student's t-test for matching tree-ring patterns. <u>Bulletin of the Institute of Archaeology University of London</u> 20, pp. 101-105.

Sapwood Estimates and the Dating of Short Ring Sequences

J. Hillam,[*] R.A. Morgan[*] and I. Tyers[+]

[*] Sheffield Dendrochronology Laboratory
Department of Archaeology and Prehistory
University of Sheffield
Sheffield S10 2TN

[+] Dept. of Greater London Archaeology
Museum of London
London Wall
London EC2 5HN

ABSTRACT

A general view of some of the problems involved in the analysis of archaeological timbers (see Hillam this volume) has prompted us to look in more detail at two aspects of tree-ring dating: sapwood estimations in oak (Quercus spp), and the use of short ring sequences. These are either not encountered or are not important in studies of modern trees or in the examination of timbers for chronology-building.

We present sapwood statistics for different regions of north-west Europe. The number of sapwood rings varies with age, depending on whether the tree is mature or immature. For immature trees, average ring width is also related, although it does not affect the sapwood number of mature trees. We recommend the use of a range of 10-55 sapwood rings for all oak samples from the British Isles. Geographical variation in sapwood number means that these values have to be modified for imported timbers.

We also examine the possibilities of dating ring sequences with as few as 30-50 rings. Although short ring sequences are often used unwisely, we conclude that under certain conditions they can be used for relative, and occasionally, absolute dating. Some examples of the use and mis-use of such patterns are given.

INTRODUCTION

A general view of tree-ring dating in relation to archaeological timbers is given elsewhere by Hillam in this volume. It is apparent from that study that there are two aspects of tree-ring dating which may be abused. These are the estimation of the number of oak sapwood rings, and the use of short ring sequences, both of which emphasise the difference between archaeological dating and studies on modern trees or timbers for chronology-building. Sapwood is only important if a precise felling date is required for archaeological interpretation, and timbers with long ring sequences are selected for chronology-building, so the problems of trying to crossmatch short ring sequences do not arise.

Previous sapwood estimates have mostly been derived from timbers with more than 100 rings. Since such timbers are not common from English archaeological sites, it is also necessary to include sapwood data from timbers with fewer rings. By reviewing published figures and collecting data from new samples, we present a working model for sapwood estimation. Such a model can be used consistently until more data are collected, and more rigorous statistical tests carried out.

Table 1. Statistics for a number of different sapwood data-sets. The Iron-age and Roman data sets are the authors unpublished data. The Irish, English and French data from mostly modern trees are unpublished data from Drs. Mike Baillie and Jon Pilcher. The Finnish material is unpublished data from Dr. Keith Briffa. The method of calculating 95% confidence limits follows that of Hughes et al 1981.

Geographic Area	Description	Sample Number	Absolute Range	Arithmetic Mean	Standard Deviation	Arithmetic Skewness	Log. 95% Range	Log. Skewness
Ireland	Baillie 1982 fig 2;3	65	14 to 62	31.32	8.99	0.420	16.74 to 53.93	0.088
	Modern & Post Medieval	49	14 to 52	32.24	8.91	0.001	16.85 to 56.65	0.489
England	Roman	61	10 to 66	23.59	9.71	2.40	10.95 to 47.55	0.001
	Iron Age Fiskerton	58	9 to 61	23.47	11.71	0.815	7.90 to 55.00	0.032
	Modern Norwich & Ludlow	30	16 to 62	29.43	10.33	1.464	14.44 to 53.69	0.082
	Hughes et al 1981	175	10 to 55	25.8	8.0	0.95	13.7 to 44.6	0.005
France	Modern	118	12 to 49	26.58	7.04	0.591	15.25 to 43.26	8.1×10^{-7}
Germany	Hollstein 1980 amended	446	7 to 66	19.00	7.54	1.707	8.22 to 37.95	4.3×10^{-5}
Sweden	Brathen 1982	69	9 to 32	15.84	4.65	1.034	8.73 to 26.55	0.097
Finland	Modern	60	7 to 24	13.85	3.19	0.051	8.32 to 21.80	0.239

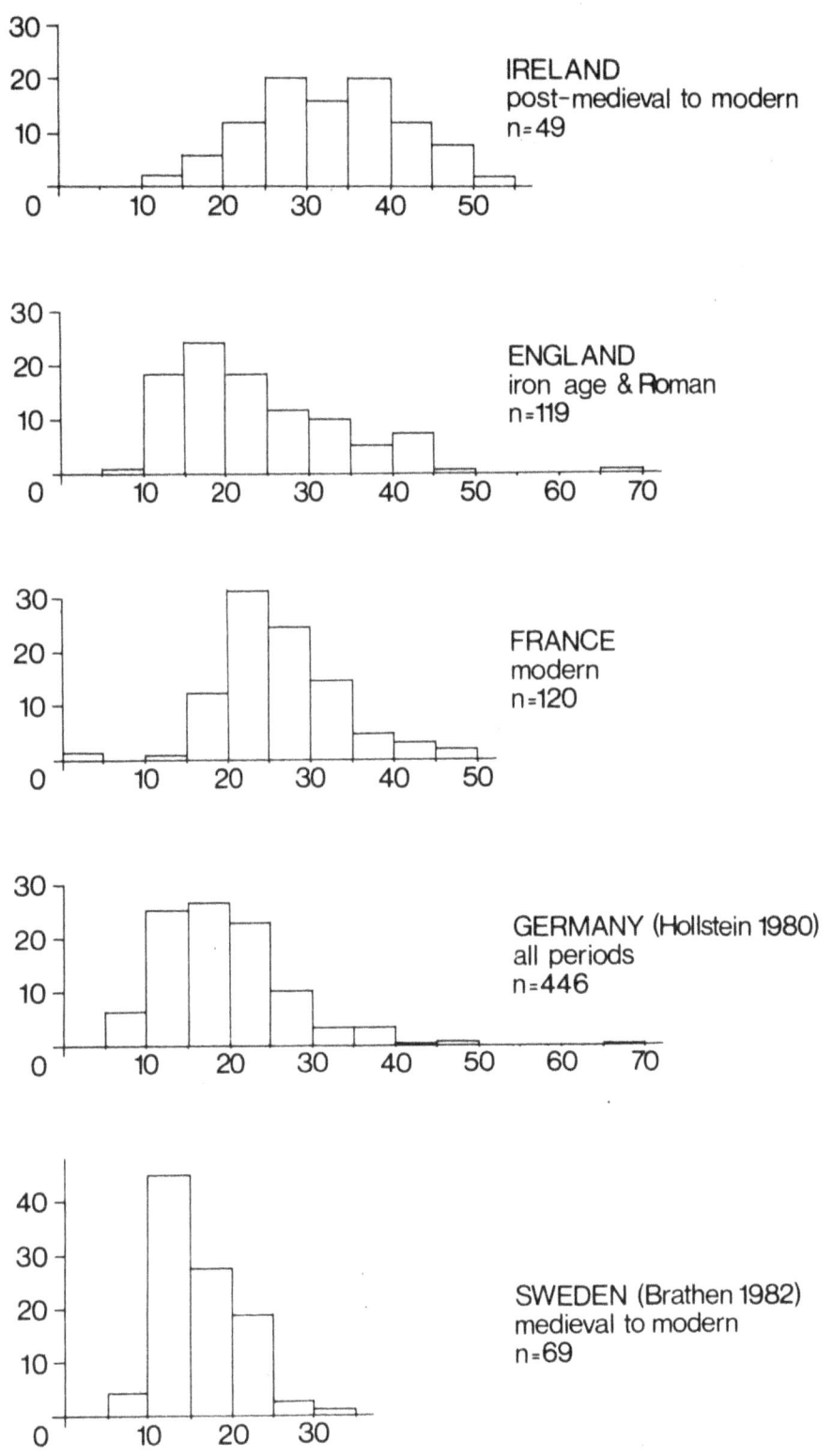

All graphs: Horizontal axis - number of sapwood rings
Vertical axis - number of samples (percentage)

Figure 1. Histograms of sapwood data. Note the marked skewness of most sets, the number of extreme values, and the general reduction in mean and variance eastwards across Europe. See Table 1 for details of the sources of data.

Because archaeological sites often offer little opportunity to select the most suitable samples, the examination of timbers with less than 100 rings is routine, and the examination of those with 30-50 rings is becoming more common. We look at the problems of crossmatching short sequences, and discuss the reliability of tree-ring dates deduced from them.

SAPWOOD

Although dendrochronological dating allows the last ring of a timber to be dated precisely, the absence of some or all of the sapwood on a tree-ring sample can cause difficulties for the estimation of its felling date. This problem is common due to the less durable properties of sapwood in oak trees compared to the heartwood; it is often absent as a result of woodworking techniques, uncountable because of woodworm damage, or lost through decay in archaeological samples. It is important to have an estimate of the number of rings likely to be absent, but the issue is complicated because the number of sapwood rings present in oak trees is highly variable, and difficult to quantify. The number in a single tree varies both around its circumference (e.g. Hillam 1985a Figure 4) and up the length of its trunk (Hughes et al 1981); the differences between individual trees can be even larger, and there is also increasing evidence of geographical variation.

The production of a practical and reliable sapwood allowance involves counting the number of sapwood rings present on a large number of samples which extend to the bark edge, and calculating from these data a range which can then be applied to samples lacking sapwood. However, the variation is large and is not evenly distributed around a mean. The value most widely used in the British Isles is 32 ± 9 years, calculated from modern Irish trees and subsequently applied to Irish and English archaeological timbers. This value is acknowledged to be rough and ready (Baillie 1984) but it has been used generally to avoid confusion. Although the statistics are broadly correct, the value is open to criticism, firstly because $\pm$ values (standard deviations) are easily misinterpreted. 32 ± 9 indicates that 68% of samples have 23-41 sapwood rings, and 95% have 14-50 rings; for example, the year 1232 AD $\pm$ 9 is the same as 1223-1241, but 32% of the samples are likely to fall outside these limits. In addition, since the data are skewed (Table 1; Figure 1), the values are not adequately describing the data. An alternative approach is to transform the data by logarithms to counteract the skewness (Table 1), calculate the standard deviations, and present them as 95% confidence ranges. Hughes et al (1981) used this method on data from 175 samples from North Wales and north-west England, and calculated a 95% range of 13.7 to 44.6 rings. These data came from cores, largely from mature trees, taken from about 1.4m above ground. Hughes et al also carried out a detailed study which showed the variation in number of sapwood rings up the trunks of 11 trees. They concluded that the variation in a group of samples taken from a single level within a set of trees is an underestimate of the variation in a group of samples taken at random heights up the trunks. This is the more common archaeological situation. Table 1 shows how log transformations reduce the skewness of most data sets (95% log transformed confidence limits are used throughout this paper).

Whilst this variation in the number of sapwood rings is adequate for many applications, such a range can make the separation in date of a series of archaeological features difficult (see Hillam this volume), and so further studies have been made to improve the prediction of sapwood number.

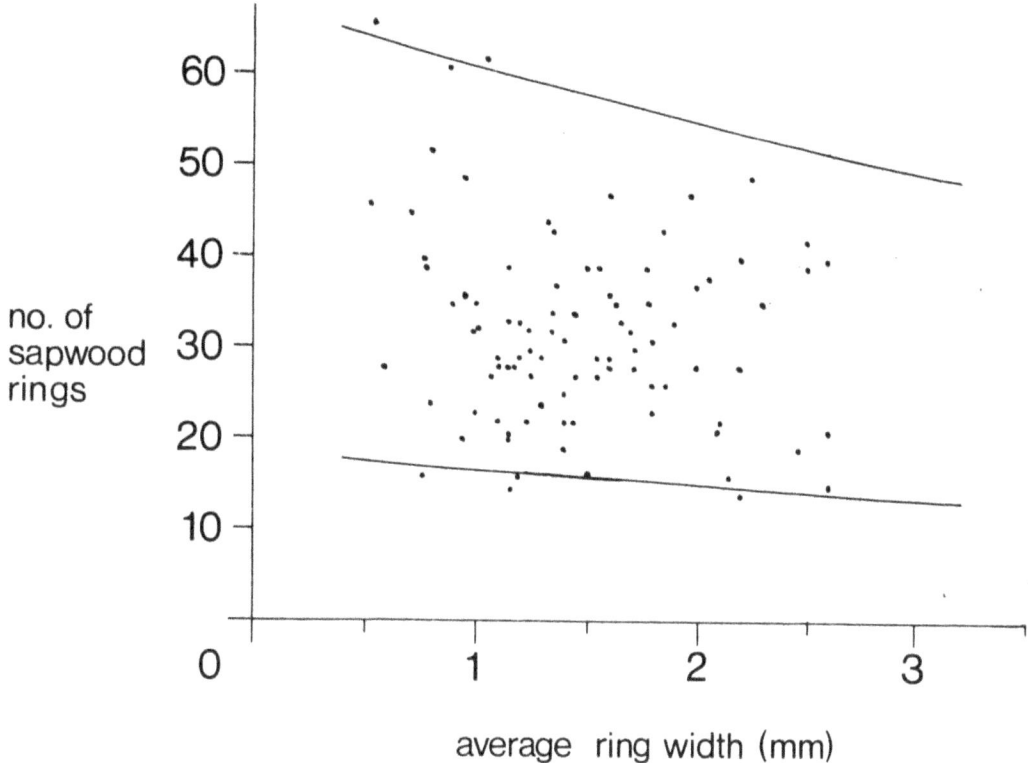

Figure 2. Relationship between sapwood estimates and average ring-widths for mature trees (over 100 years). n = 91

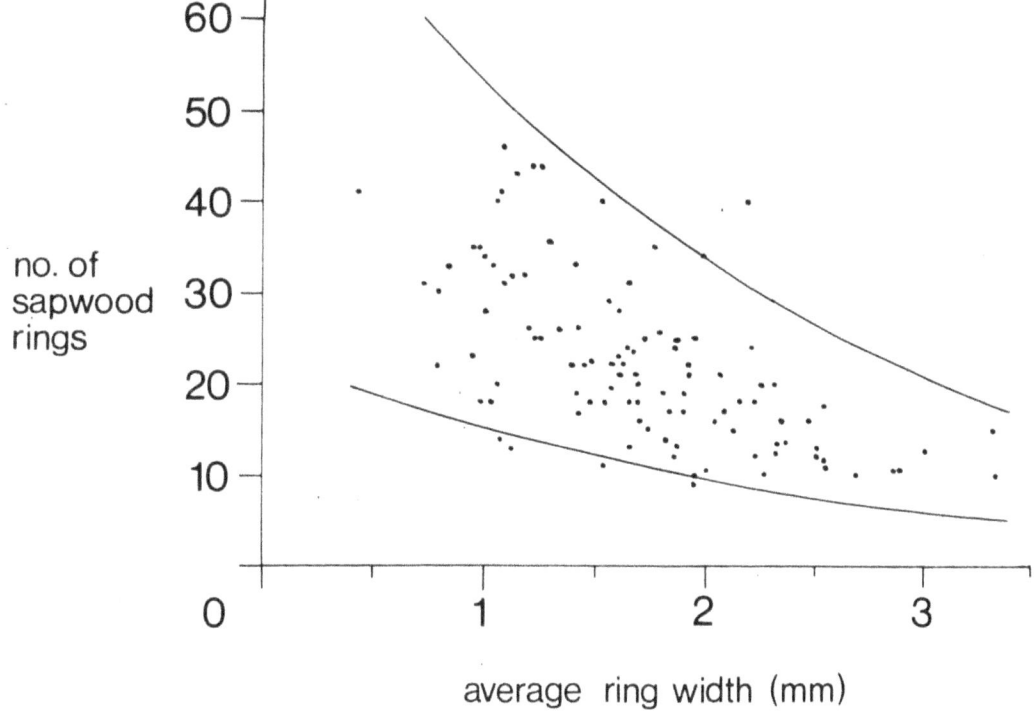

Figure 3. Relationship between sapwood estimates and average ring-widths for immature trees (less than 100 years). n = 106

This involves examining other variables (tree age, average ring width, and geographical location), and investigating their relationship to sapwood numbers.

1. Tree age

Relating tree age to sapwood number on whole sections of timber, i.e. those extending from pith to bark, is relatively simple. However many archaeological samples are only part of the cross-section of a tree trunk, and the tree's age is unknown. It is therefore difficult to collect sufficient data on the range of sapwood number compared to tree age from this source.

In addition, the inclusion of data from immature trees can introduce further problems. Because trees of less than 30 years of age are unsuitable for dating, sapwood estimates which include such data may be unrepresentative for archaeological timbers. In Germany, Hollstein (1980 Figure 21) related tree age to sapwood number for 493 data points, with tree age varying between 15 and 400 years. 47 of his samples had less than 30 rings total, including 4 with no heartwood at all. These samples must produce an underestimate of the real sapwood number. While 47 points are only 9.5% of the data, the calculated relationship is significantly affected by small changes in the data. In the British Isles, several scatter diagrams showing the relationship of tree age to sapwood number have been published; they are mostly based on mature trees and the predictive value is acknowledged to be low (Hughes et al 1981, Figure 5; Baillie 1982, Figure 2:3a).

The relationship has been examined from a different angle here. Instead of calculating regression lines from the known tree age of many samples, the English (Table 1) and German (Hollstein 1980) data were split at the arbitrary point of 100 years total age. Splitting the data at this point allows more samples to be used since even part sections of timbers can usually be classified as either 'mature' or 'immature' (this is not a strictly accurate use of the terms but it will suffice to distinguish the types). Although this is an arbitrary division, Table 2 demonstrates its value. The data in Table 1 could be used to suggest that English and Irish timbers have differing sapwood numbers; resorting the data by age shows that these differences are the effect of sampling (Table 2). The English data are mostly from 'immature' trees, while the Irish material is mostly 'mature'. Grouping the data sets in this way produces 95% ranges that can be applied to native material from England, Ireland and Wales of 15-60 sapwood rings for 'mature' and 10-45 for 'immature' trees. This division is paralleled in Hollstein's data. Providing a crude division between immature and mature timber can be made (see below), the distinction is useful for dating purposes. Collecting more data may allow better divisions to be made.

2. Average ring width

A second variable is the average ring width of the sample which is determined routinely. The actual value to be used needs some thought. Many trees exhibit changes in average ring width from the centre to the bark. Should the average used be the total average width; the heartwood average only (in archaeological samples the softer sapwood may have

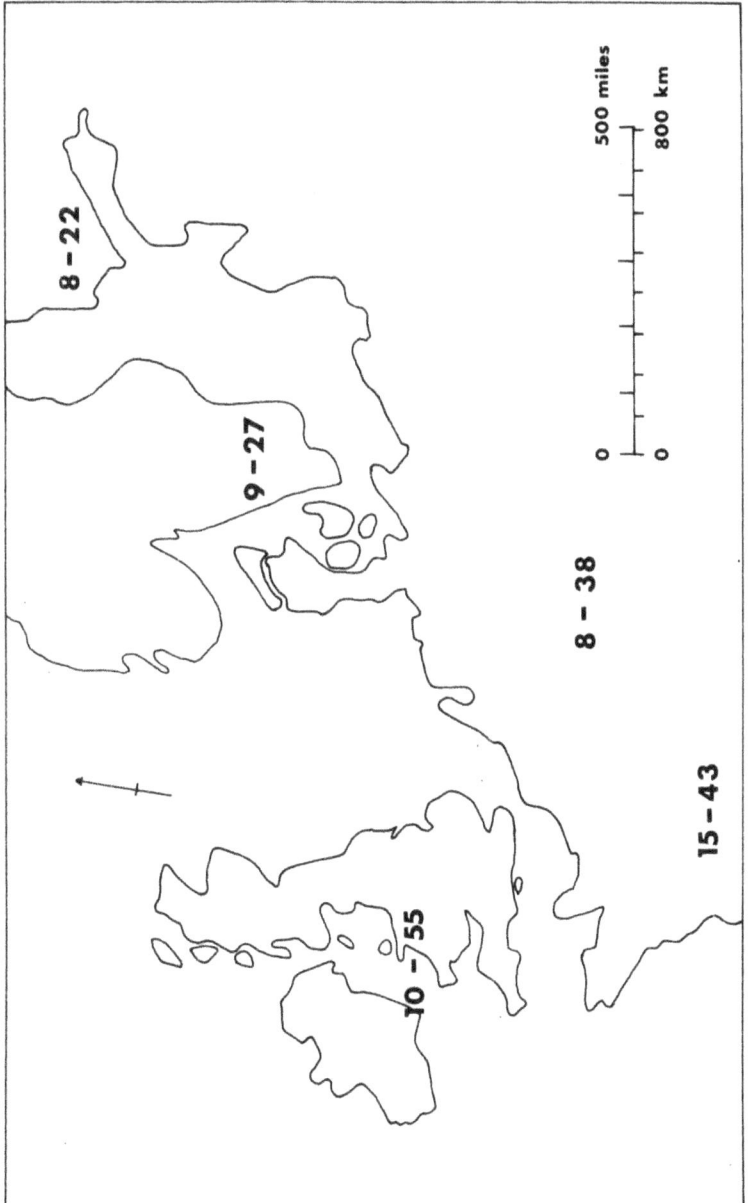

Figure 4. Geographical variation in sapwood data. Swedish values are based on mostly immature trees, Finnish on mature trees. The remainder include data from both age classes. Details in Table 1.

Table 2. Statistics showing the differences between 'mature' and 'immature' data for the British Isles and Germany.

Geographic Area	Age type	Sample Number	Absolute Range	Arithmetic Mean	Standard Deviation	Arithmetic Skewness	Log. 95% range	Log. Skewness
British Isles	Mature	91	14 to 66	31.74	10.23	0.450	15.59 to 58.15	0.067
	Immature	106	9 to 46	22.35	9.11	0.550	9.16 to 46.33	1.3×10^{-4}
Germany	Mature	214	11 to 66	22.83	7.68	2.20	11.63 to 40.52	0.085
	Immature	232	7 to 35	15.47	5.36	0.668	7.39 to 28.83	0.002

contracted through drying out or from physical pressure); or the average of the last few heartwood rings before the heartwood/sapwood transition, and if so, how many? Previous work in Britain on mature trees has suggested that the trends are not very useful for sapwood prediction (Hughes et al 1981, Figure 5; Baillie 1982, Figure 2:3b). However, two recent papers have suggested that sapwood number does vary with differing average ring widths (Fletcher 1982; Fletcher & Tapper 1982). The main difference between the examples of Hughes and Baillie already cited and Fletcher's proposal is that the latter's values are based on trees under 100 years old when felled.

Using the 'mature' and 'immature' data sets (Table 2), scatter diagrams were produced using sapwood number and total average ring width. The overall British Isles data are presented here (Figures 2, 3). The individual sets only confirm the similarity between English and Irish sets of similar age.

For trees over 100 years old when felled (Figure 2), there is a slight relationship with average ring width, but it is of little predictive value. 95% confidence limits of 15-60 rings can safely be applied to presumed native timbers without any reference to average ring width.

The data for trees between 30 and 100 years old when felled are significantly different (Figure 3). A range of 10-45 years can be applied, and the variation with average ring width is of predictive value. However, in practical terms, this relationship is of little use (see below), so the graph should be used with some care.

Although Fletcher's results (Fletcher 1982) are similar, his figures do not adequately allow for the variation of these data sets, and are unsuitable for use with English archaeological material. Problems of applying them to archaeological phases are discussed in Hillam et al (1984).

3. Geographical variation

Using the data in Tables 1 and 2, the geographical variation of sapwood number can be demonstrated (Figures 1, 4). While there are many gaps in the areas sampled, an east-west trend is apparent and is statistically significant. There may also be a north-south trend: Scottish samples appear to have more sapwood rings than other British Isles material (Baillie pers. comm.). The cause of such a variation is unknown but may be related to the climatic Atlantic-Continental trend. Geographical variation is a potential problem in the interpretation of tree-ring dates, since the application of sapwood allowances to imported timbers (often of unknown origin) would be invalid; see Baillie et al (1985) for a discussion of the problem.

In summary, there is a slight variation in the number of sapwood rings according to tree age, and the sapwood width is also related to average ring width for trees younger than about 100 years old. Geographical location is also important, with the number of sapwood rings increasing from east to west across north-west Europe. However owing to the difficulty in determining tree age and provenance in archaeological material, we recommend the use of a range of 10-55 sapwood rings for all samples. This will result in a maximum felling date range of 45 years where there are only a few remaining sapwood rings. The range will be reduced when more sapwood rings are preserved.

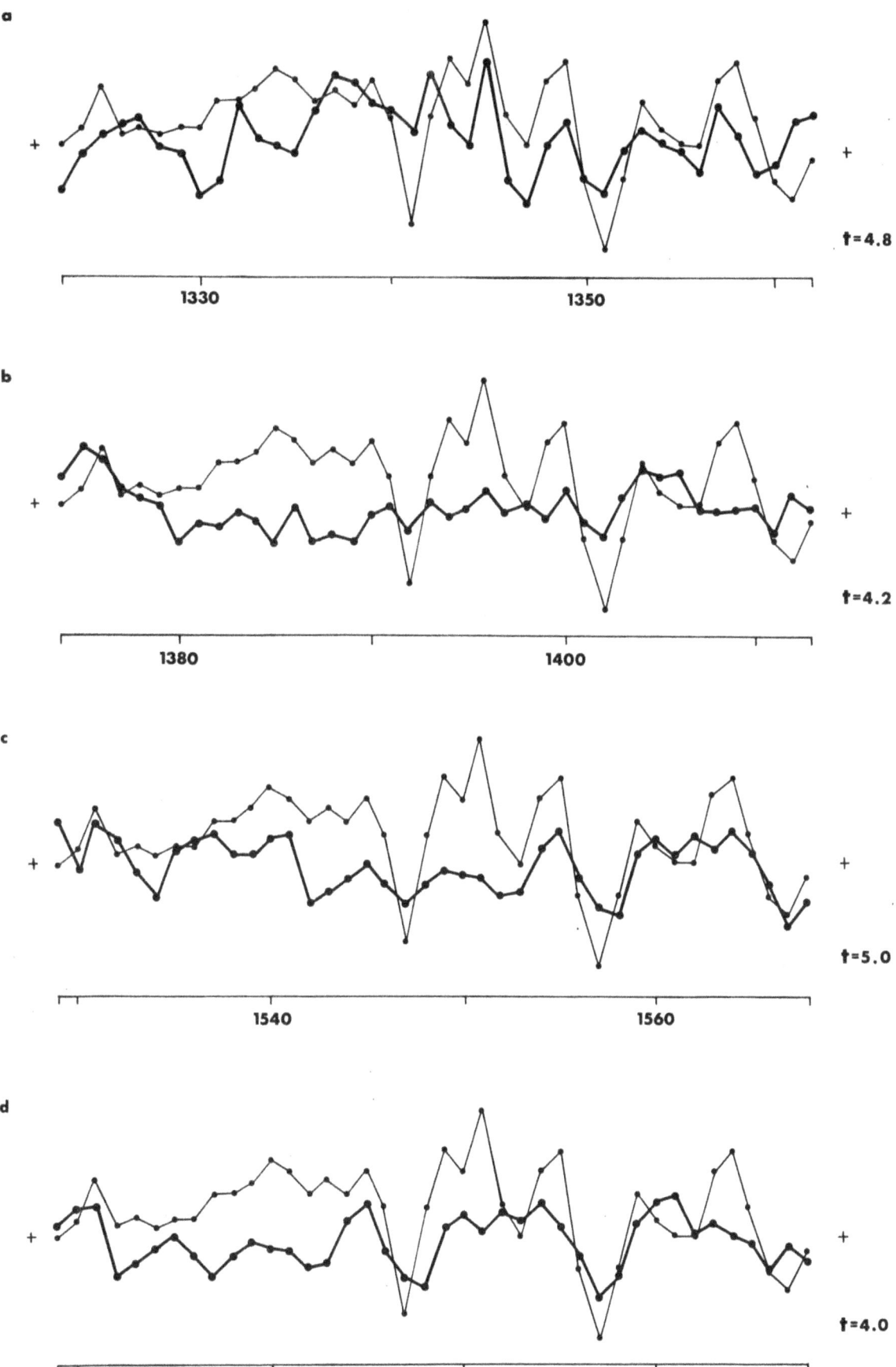

Figure 5. A 40 year section of the Bishops' House, Sheffield, master chronology (thin line) compared to the Yorkshire master curve (bold line). All four positions of agreement give high t-values and reasonably acceptable visual matches. Position (c) is the correct one, but this is not obvious from the tree-rings.

DATING SHORT SEQUENCES

Hitherto dendrochronology laboratories have always avoided or turned away wood samples with less than 50-80 rings (the minimum number depends on the laboratory - see, for example, Baillie 1982, p. 68; Huber & Giertz 1970). This is because the early development of European tree-ring dating in the 1970's was concerned largely with chronology-building: that is, cross-matching the longest available series of rings into chronologies which would extend as far back in time as possible. Only through concentrating on trees of at least 200-300 years of age has it been possible to construct a chronology back to 5289 BC for Northern Ireland, and at the same time ensure the reliability of that chronology (Pilcher et al 1984).

Much of the chronology-building work involved the use of oaks killed naturally by peat bog development or flooding by rivers. However, during research in the late 1970's, much archaeological material became available with the expansion of rescue archaeology in both rural and urban contexts. Some proved of value for chronology-building: for example, the London waterfronts (e.g. Bateman & Milne 1983) or Viking York (Hall 1984); but carpenters tended to make use of younger trees for practical and economic reasons. Much of the archaeological record therefore consists of planks, posts and beams from trees less than 100 years old, in addition to large quantities of small roundwood of various wood species which was used for fencing or infill such as wattling.

We have always aimed at extracting as much information as possible from the relatively small surviving remnants of this important raw material. Samples have been collected and analysed from a wide variety of material in terms of size, age and species, to explore the potential of each for tree-ring studies. Where ring sequences from immature trees are concerned, several points need to be clarified.

First, absolute dating in the British Isles is usually possible only on oak wood; no other species are found with sufficient frequency to enable long chronologies to be built, nor has it yet been proved that they can be dated with certainty by means of the oak chronology. Usually a site chronology, made by averaging the ring widths from matching sequences from individual oak timbers, needs to span at least 100 years to stand any chance of reliable absolute dating (see Hillam this volume).

Second, absolute dating, while a priority in many cases, is not always the sole aim of the exercise. The archaeologist may be equally interested in resolving the relationship between timbers on a complex multi-phase site. For this purpose, very long ring series would be less important than defining the year in which the individual trees were felled (whether relative or absolute). The presence of the outer part of the tree is necessary for this and untrimmed roundwood posts prove to be ideal material. Unfortunately they also tend to have less than 100 growth rings.

From some sites, a large proportion of the wood contains less than 50 rings, for example, Neolithic Auvernier-Port in Switzerland 83% (Orcel & Egger 1982); Bronze Age Meare Heath track, Somerset 44% (Morgan 1982); Iron Age Fiskerton, Lincolnshire 61% (Hillam unpubl); Viking Coppergate in York 33% (Hillam unpubl). If these samples were not examined, a potential source of fine detail, both within and between structures, might be lost. This assumes that such samples are intrinsically datable, wherein lies the main criticism of work with short sequences. It has been claimed that

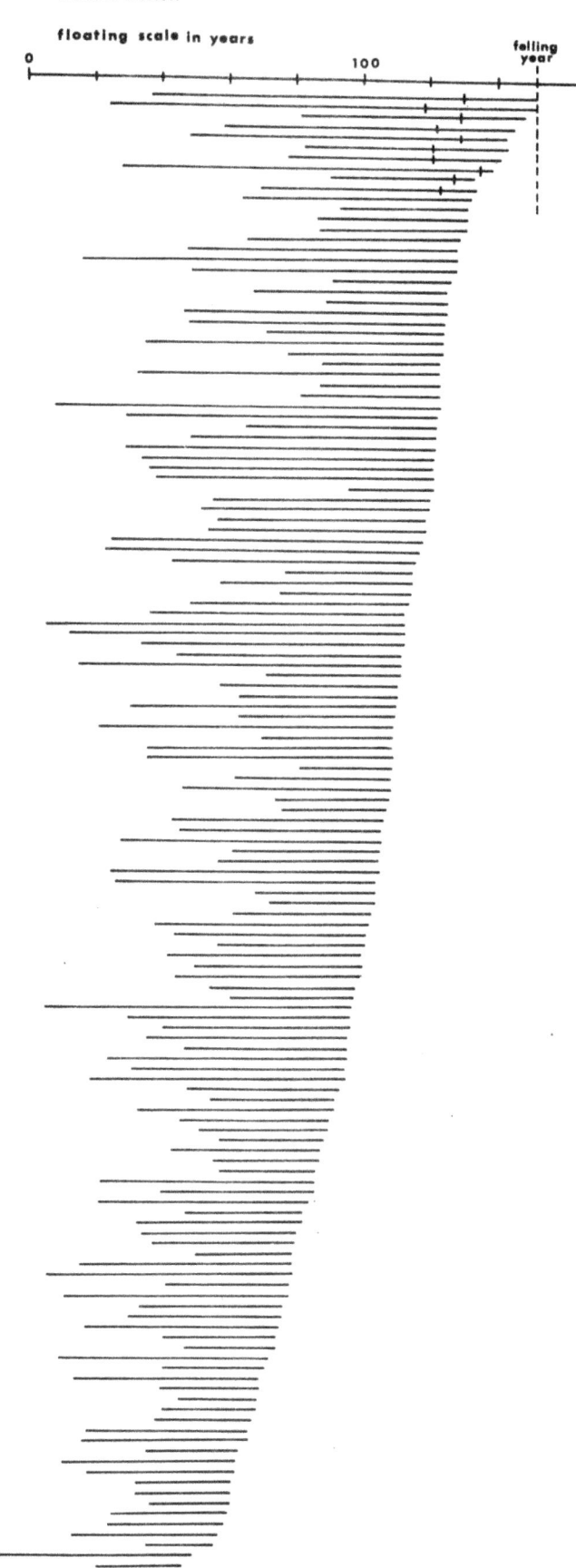

Figure 6. Typical bar diagram from a prehistoric trackway in Somerset, in this case the LBA Meare Heath track. Each bar represents the relative span in years of the cross-matched rings of each piece of wood. Most of the ring sequences are short (44% under 50 years, 93% under 100 years). Despite the cross-matching of 145 curves, the chronology still only spans 152 years. Many of these timbers probably originated in the same trees - groups of timbers probably from the same individual tree could be recognised along the track's length.

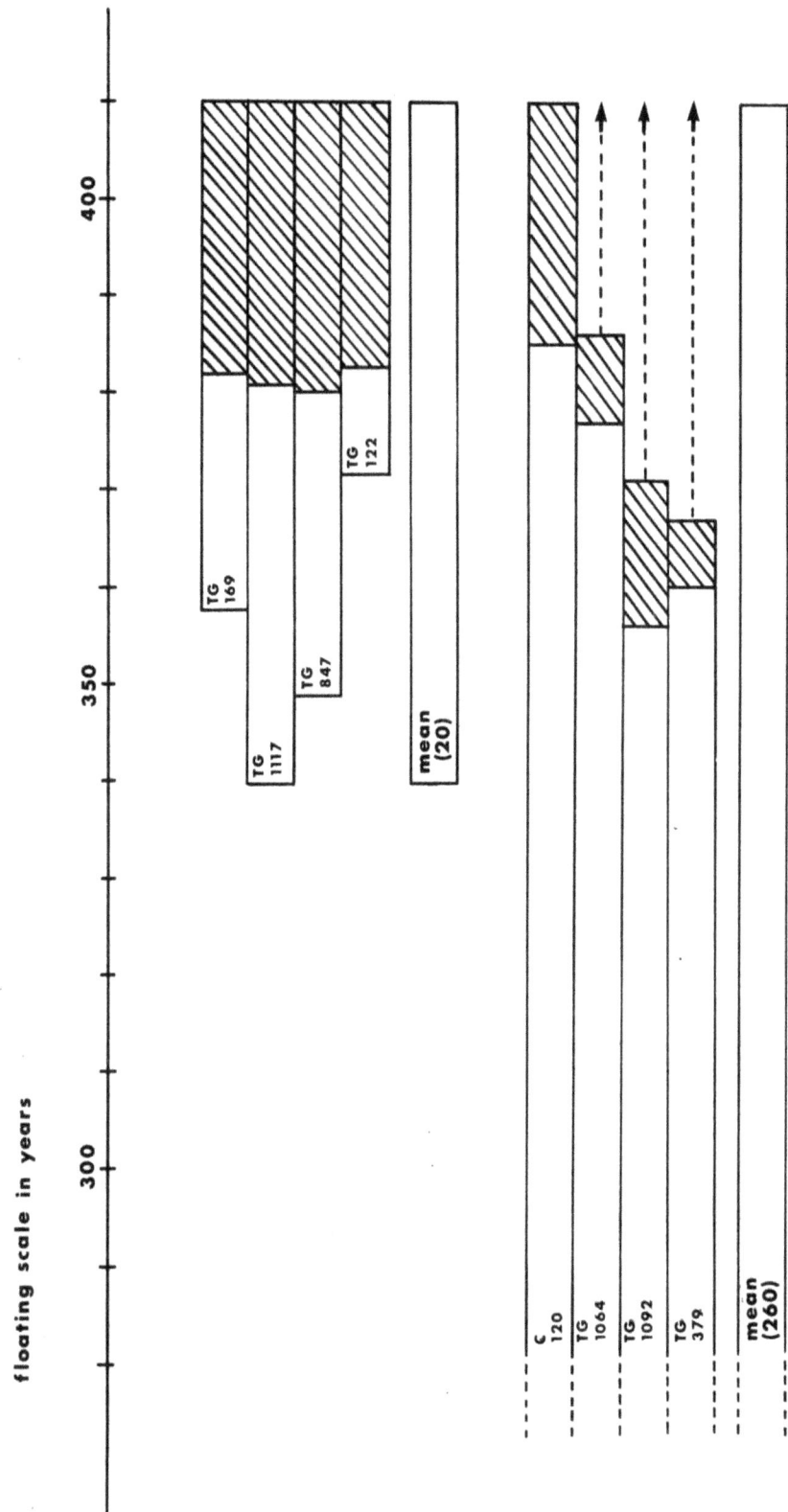

Figure 7. Example of how short sequences from immature roundwood with bark edge can assist in defining the felling year. Here planks from the neolithic Sweet track in Somerset were mostly trimmed of sapwood, leaving traces on only 5 out of 260 planks from which the rings were cross-matched. One, SWC120, was suspected to extend to the bark edge, but no confirmation could be found on the other planks. The felling date was finally proved by cross-matching a 71 year mean curve based on 20 young oak posts and pegs from farther up the track. It was then known that the trees for constructing the track were felled in arbitrary year 410, which may in the future be equated to a calendar year in the early 4th millennium BC. Hatching represents sapwood.

ring patterns of less than 50 years may not be unique, i.e. that the pattern could be repeated over more than one period of time (Huber & Giertz 1970). Work on series from modern trees shows that this is certainly true for those with less than about 30 rings (Hillam unpubl). The problem of duplication for samples with 30-50 rings is lessened however if the ring pattern is replicated many times over, by extensive sampling from the same structure. The actual number of rings therefore is less crucial to relative dating than the number of related samples. When only one or two samples are examined, it is possible to find several synchronous positions, all with good t-values to support them, and no one match stands out better than the others (Figure 5). It is difficult to rely on the t-values alone, since the computer program which calculates them (Baillie & Pilcher 1973) was designed for use with ring sequences of 80 years or over. However if the short curve can be matched with a series of others from the same source, the possibilities of reliable crossdating are increased.

Attempts at dating short sequences

The ill-advised use of short ring series from only one or two samples can be illustrated by two examples. The felling of trees for the Roman waterfront at Custom House in London is presently 'dated' on the strength of one timber with 39 rings including sapwood, which is tenuously linked to a long dated chronology by 18 years of overlap (Fletcher 1982). In the other case, a piece of wood with only 17 rings was considered to be dated by means of a well-known signature, or characteristic pattern of rings, and documentary evidence (Siebenlist-Kerner 1983).

A more acceptable application is exemplified by work in Switzerland (e.g. Egger 1983) which illustrates the refinements of tree-ring dating, particularly in relative dating: definition of house plans at different periods, the duration of settlement, repairs to structures and so on. One chronology from Auvernier is represented by over 150 samples, yet it covers only 70 years. However concern is felt about the use of ring sequences under 30 years (e.g. Orcel & Egger 1982).

On the neolithic site of Alvastra in southern Sweden (Bartholin 1978, 1983), a floating chronology of just over 70 years was constructed from oak ring patterns. The felling dates showed an 18 year period of intensive, almost annual, building activity, a 20-25 year abandonment and a little activity in years 40-42 on the relative time scale. Almost all the piles were from trees 20-50 years old, yet the information they gave was of vital importance to the interpretation of the site.

Short sequences have always played a major role in the relative dating and interpretation of the prehistoric trackways in the Somerset Levels. Many were constructed of roundwood rods and poles, of wood species such as hazel (Corylus) and ash (Fraxinus) which offer information along the lines of the Swiss and Swedish work (e.g. Morgan 1984). The trackways which include oak were made not only of large planks but also small slats and stakes, and it is important to establish their temporal relationship to the planks.

For example, the LBA Meare Heath track (Coles & Orme 1978) has provided over 140 pieces of wood, the patterns from which have been cross-matched into a chronology only 152 years long. Their origin in a few trees is suggested by the good quality of matching and by the resulting bar diagram

Table 3. Fiskerton t-values. The fourteen sequences have between 29 and 39 rings.

	49	97	108	116	122	131	137	143	149	157	205	255	335	342	Master
49	-	**5.3**	3.2	0.2	**3.8**	**3.8**	2.8	**3.6**	**4.8**	3.3	**4.9**	1.9	**5.4**	**4.8**	**4.7**
97		-	**3.5**	0.2	**5.8**	2.2	**4.6**	2.9	**5.0**	3.2	**3.5**	1.0	**3.5**	**6.0**	**4.3**
108			-	**3.7**	3.2	1.9	**4.3**	**4.2**	**4.2**	3.0	2.2	2.9	1.7	2.4	**4.4**
116				-	**9.7**	2.3	**3.7**	**4.6**	**5.6**	**5.0**	**4.0**	**6.4**	**5.0**	**5.5**	**5.0**
122					-	2.2	**4.9**	3.4	**4.3**	**5.2**	2.8	1.2	**4.6**	**5.3**	**4.4**
131						-	3.0	1.4	3.0	1.4	3.2	0.8	3.4	1.5	2.8
137							-	2.6	2.6	**3.5**	1.8	0.9	2.5	**3.5**	**4.8**
143								-	**3.8**	**3.6**	2.6	1.5	2.1	1.9	3.4
149									-	2.9	**3.5**	**4.9**	**4.0**	3.1	3.2
157										-	3.2	**6.4**	**5.2**	**3.8**	**4.0**
205											-	**3.9**	**7.3**	3.4	**5.3**
255												-	1.9	**3.9**	**3.6**
335													-	**5.5**	**3.8**
342														-	**5.0**

Table 4. Results of comparisons between 14 Fiskerton ring sequences with less than 40 rings and the Fiskerton master (in which they are not included)

Sample number	t-values over 3.5	correct t	incorrect t
49	1	4.7	-
97	1	4.3	-
108	1	4.4	-
116	1	5.0	-
122	2	4.4	3.6
131	-	-	-
137	1	4.8	-
143	-	-	-
149	1	-	3.6
157	2	4.0	3.9
205	1	5.3	-
255	2	3.6	4.0
335	1	3.8	-
342	2	5.0	3.8

illustrating the relative position of each ring series (Figure 6). This shows a gradual fall-off in end years of the timbers, now found to be typical of the Somerset oak assemblages. It implies that many planks and stakes could be split from a few trees. Familiarity with the overall growth pattern resulting from the analysis of such quantities of material enables short ring series such as those from the split oak stakes to be matched with confidence (Morgan 1982). Sapwood was present on a few timbers and will give an exact felling year in the event of absolute dating of this as yet floating chronology.

In other cases, short series have been valuable in defining the end of the chronology and the cutting year of the trees. A series of oak roundwood pieces from the neolithic Sweet track (site TG; Coles & Orme 1984) with 20-70 rings were cross-matched into a mean curve which extended out to the bark surface, and hence indicated the year of felling. This in turn matched well (Figure 7) with the only plank retaining all or most of its sapwood from the other end of the track (site C). This confirmed the otherwise uncertain felling year for all the oak trees used in the track, and potentially its construction date (Morgan 1984, 53).

The trackways have the advantage of being largely single period structures with little evidence of later repairs or alterations. This contrasts with Fiskerton in Lincolnshire, where the remains of an Iron Age structure, possibly a causeway, were excavated. Most of the 180 posts were sampled, and because many were untrimmed roundwood, the relative dating was often exact to the year. The results (Figure 8) showed that the causeway had a complicated life of construction and regular repairs which lasted at least 150 years (Hillam 1985b). Many of the dated ring sequences were short, in particular those from the fourteen timbers felled in the year 100 on the arbitrary scale, which have 29-39 rings. Their dating is reliable because they match each other as well as matching the master curve (Table 3). However, if the sequences are compared with the master alone, the results are not so conclusive and some would be undatable (Table 4). Figure 9 shows the ring pattern of sample 157 which appears to match in two places along the master. If this sample was the only one available for analysis, it would be considered undated, as the two cannot be separated by the visual match, the t-value or the signatures. Fortunately sample 157 matches 13 others at position B, and so can be dated. Some of the other matching sequences are illustrated in Figure 10. If the Fiskerton site master is crossdated against Irish or German reference curves (Pilcher et al 1984), then these 14 short sequences will be absolutely dated, although it should be noted that their ring width data have not been included in the site master. Absolute dating therefore depends upon the availability of longer ring sequences from the same site.

The examples shown in Figures 5 and 9 amply indicate the dangers of attempting to match short ring series. The need for very thorough sampling cannot be over-emphasised if mis-matching is to be avoided. To attempt to date a structure on the basis of one or even two short sequences (e.g. Fletcher 1982, Figure 1) is a dangerous and misleading process. Even if one or two common patterns can be recognised in support of such matching, they cannot merit the name signatures unless they appear in a large number of examples and provide the necessary replication.

FISKERTON

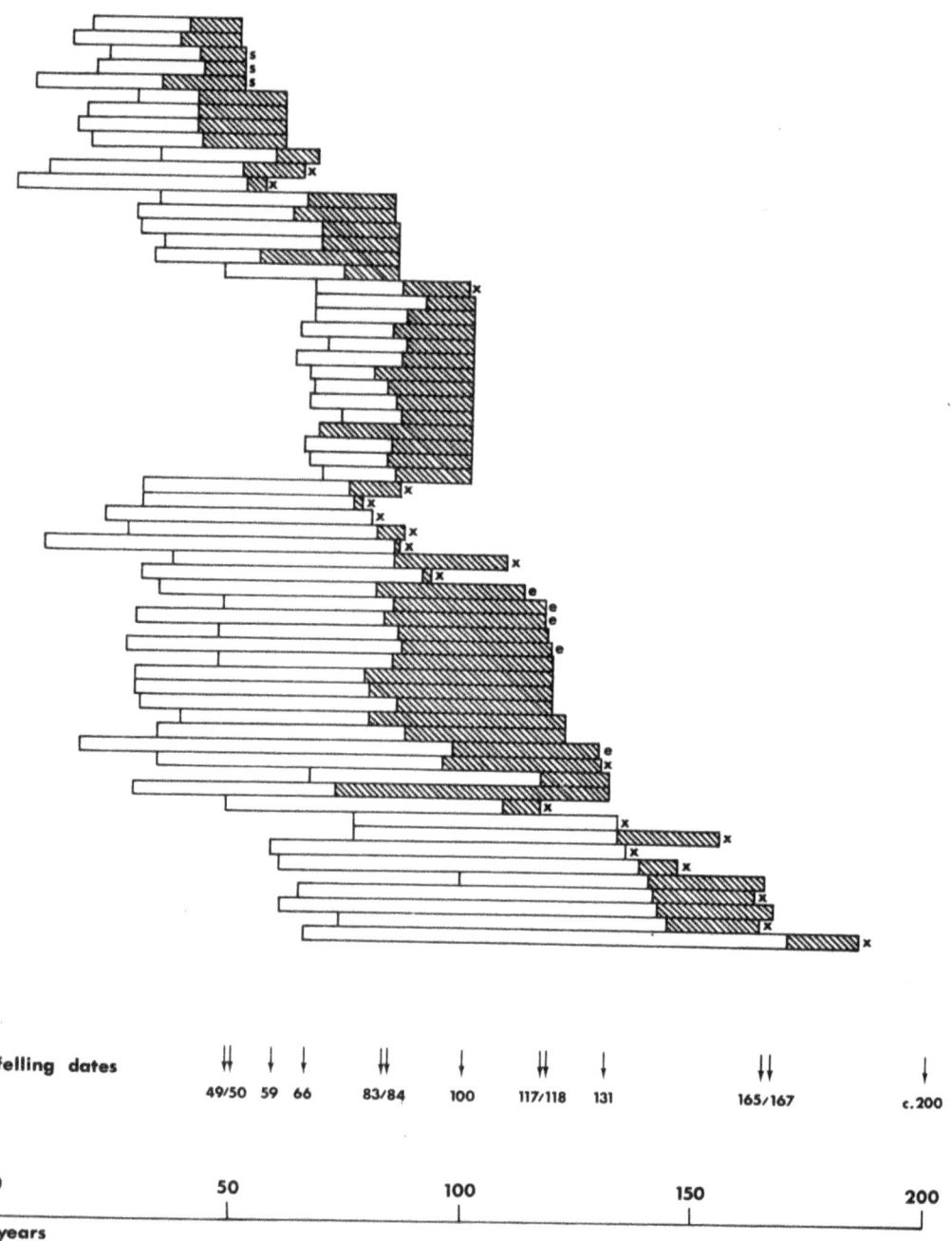

Figure 8. Fiskerton bar diagram showing the years spanned by the dated ring sequences. Open bar = heartwood rings, hatching = sapwood, s = felled in summer (all others were winter felled), x = not year of felling, e = felling date estimated by counting outer rings which were too narrow to measure with accuracy.

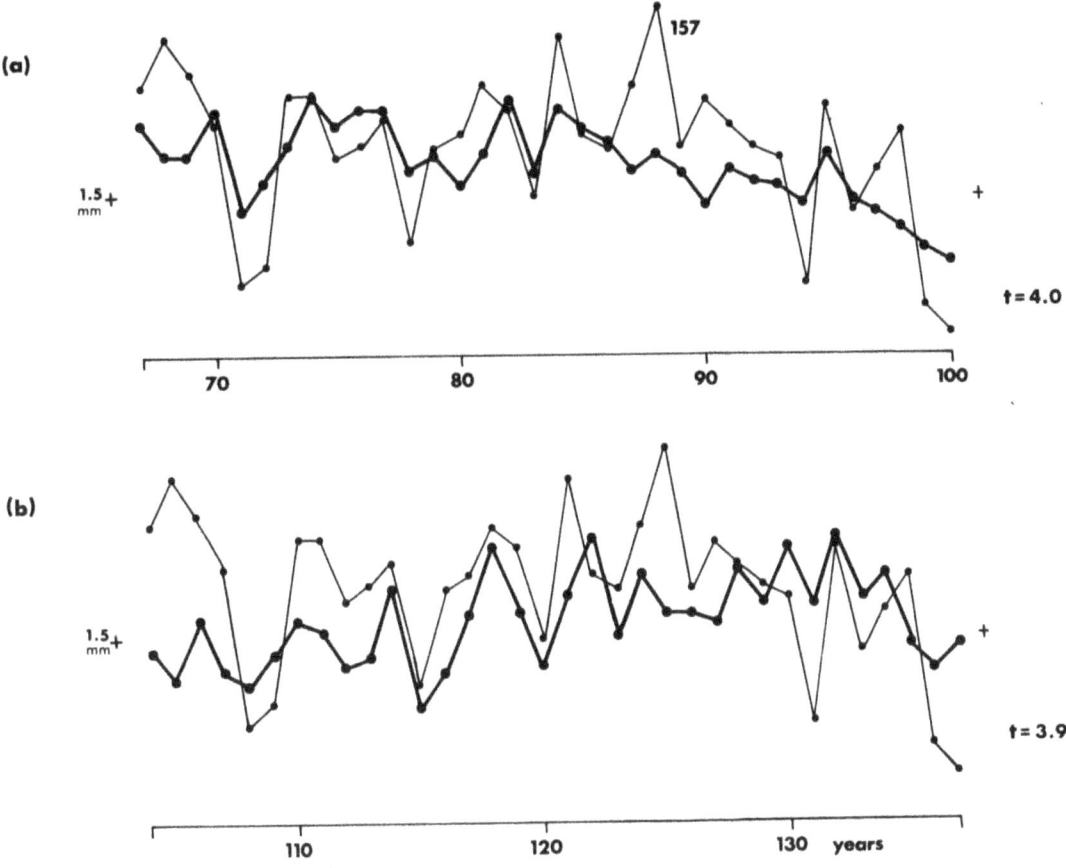

Figure 9. Fiskerton 157 (thin line) compared with the Fiskerton master curve (bold line). The two positions of synchronisation appear to be equally good. Position (a) is correct, because 157 also matches 13 other sequences, all of which end in the year 100. The vertical scale is logarithmic, the horizontal scale in years is that of the Fiskerton master curve (see also Figure 10).

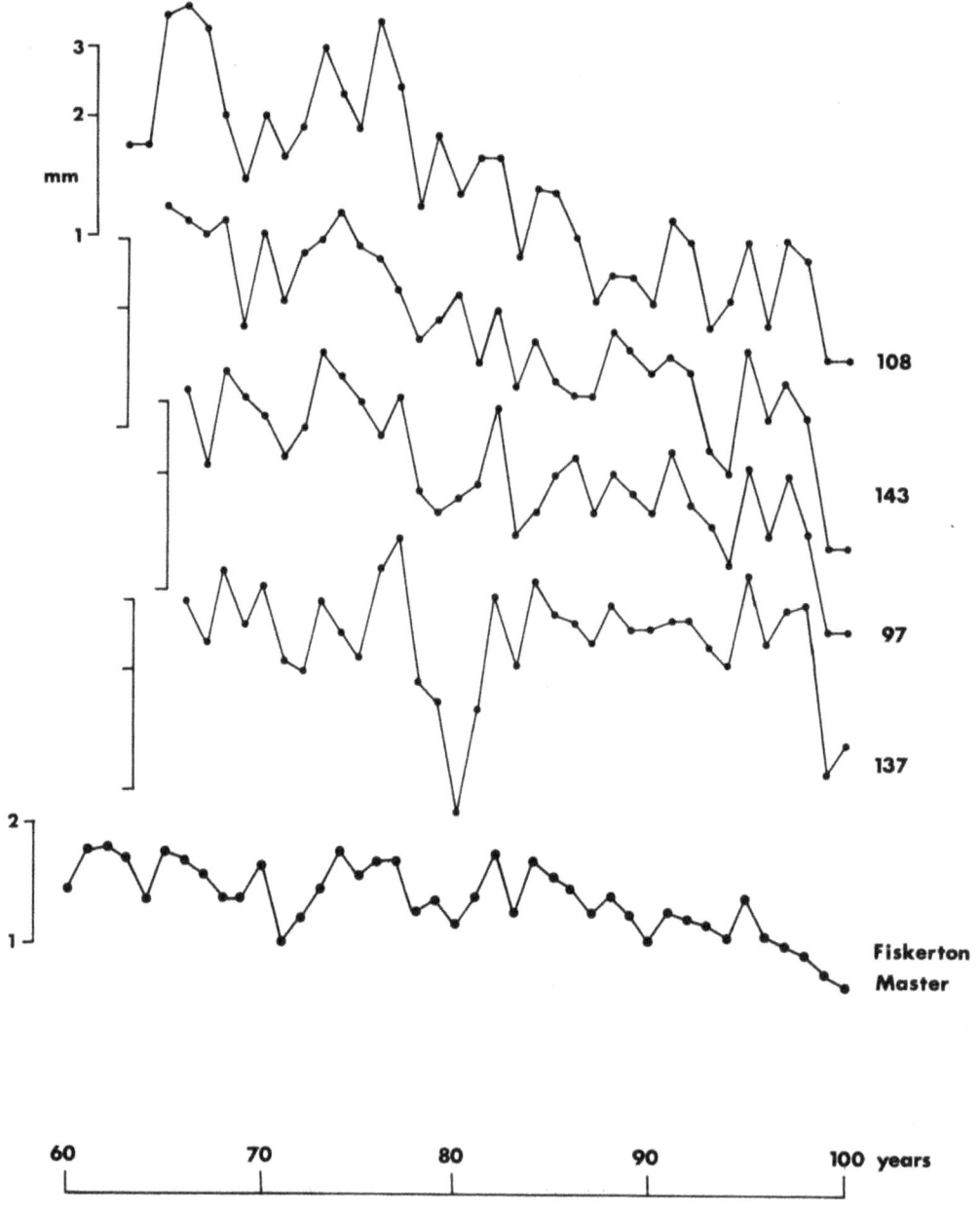

Figure 10. Matching short sequences. Four of the 14 Fiskerton sequences from timbers felled in year 100, and the corresponding section of the Fiskerton master curve.

CONCLUSIONS

The precise number of sapwood rings missing from archaeological timbers cannot be calculated. Comparing complete archaeological and modern material suggests that the main cause of variation within an area depends upon whether the tree is 'immature' (less than 100 years) or 'mature, (over 100) when felled. Average growth rate is only important if the tree is immature. There is also some variation up the trunk and around the circumference of an individual tree. Geographical variation is large and inadequately understood. If it is due to climatic factors, changes must be expected in sapwood ranges in prehistoric periods for which climatic variation is hypothesised. However, significant changes in sapwood number over the historic period are unlikely.

For material from the British Isles, the use of sapwood ranges of 15-60 years for trees over 100 years old, and 10-45 years for trees under 100, seems reliable. These ranges are 95% confidence limits and cannot allow for the extreme values sometimes encountered (Figure 1). For immature trees, careful use of average ring widths may allow smaller ranges to be used (Fig. 3). However, since tree age is often difficult to determine in archaeological samples, we suggest a general value of 10-55 years is adopted and applied to all samples native to the British Isles, (imported timbers may have different values). The continued accumulation of sapwood data from all areas should be an aim for all workers since the statistical analysis of more data may allow the 10-55 range to be further refined.

It is also concluded that short ring sequences can be cross-matched but with reservations. Long series of rings will always be easier and more reliable to date, but the nature of archaeological timbers forces us to consider all available samples. It has been found that the number of samples per structure is more important than the number of rings, and we suggest extensive sampling: 10 timbers all with 30-40 rings may be datable (at least relatively), whereas one timber with 80, or even 100, rings may not. Work on very short sequences with less than about 30 rings, or on single sequences with less than 50-80 rings, should be avoided. Further work is undoubtedly needed on short sequences, using modern samples as reference material, but the study of timbers with few rings should be encouraged in view of the refined information they provide.

Acknowledgements

The Sheffield Dendrochronology Laboratory (JH and RAM) is funded by the Historic Buildings and Monuments Commission for England. HBMC and GLC provided the funding for IT. We would like to thank Keith Briffa of the Climatic Research Unit at Norwich for unpublished sapwood data, and Mike Baillie and Jon Pilcher for sapwood data and discussion about this paper.

REFERENCES

Baillie, M.G.L. 1982. Tree-ring Dating and Archaeology. London: Croom Helm.

Baillie, M.G.L. 1984. Some thoughts on art-historical dendrochronology. Journal of Archaeological Science 11, pp. 371-393.

Baillie, M.G.L. & Pilcher, J.R. 1973. A simple crossdating program for tree-ring research. Tree Ring Bulletin 33, pp. 7-14.

Baillie, M.G.L. Hillam, J., Briffa, K.R. & Brown, D.M. 1985. Re-dating the English art-historical tree-ring chronologies. Nature 315, pp. 317-319.

Bartholin, T.S. 1978. Alvastra pile dwelling: tree studies. Fornvannen 73, pp. 213-219.

Bartholin, T.S. 1983. The combined application of dendrochronology and wood anatomy. In: Dendrochronology and Archaeology in Europe. (eds Eckstein, D., Wrobel, S. & Aniol, R.) Mitteilungen Bundesforshung. Fors-tund Holzwirtschaft 141, pp. 79-92.

Bateman, N. & Milne, G. 1983. A Roman harbour in London: excavations and observations near Pudding Lane, City of London 1979-1982. Britannia XIV, pp. 207-226.

Brathen, A. 1982. A tree-ring chronology from the western part of Sweden. Sapwood and a dating problem. PACT (Journal of the European Study Group on Physical, Chemical and Mathematical Techniques Applied to Archaeology) 7, pp. 27-35.

Coles, J.M. & Orme, B.J. 1978. The Meare Heath track. Somerset Levels Papers 4, pp. 11-39.

Coles, J.M. & Orme, B.J. 1984. Ten excavations along the Sweet track (3200 BC). Somerset Levels Papers 10, pp. 5-45.

Egger, H. 1983. Dating of Neolithic and Bronze Age sites. In: Dendrochronology and Archaeology in Europe. (eds Eckstein, D., Wrobel, S. & Aniol, R.) Mitteilungen Bundesforshung. Forst- und Holzwirtschaft 141, pp. 169-178.

Fletcher, J.M. 1982. The Waterfront of Londinium: the date of the quays at the Custom House site reassessed. Transactions of the London & Middlesex Archaeological Society 33, pp. 79-84.

Fletcher, J.M. & Tapper, M. 1982. Tree-ring dates. Vernacular Architecture 13, p. 49.

Hall, R. 1984. The Viking Dig. London: Bodley Head.

Hillam, J. 1985a. Recent tree-ring work in Sheffield. Current Archaeology 9(1), pp. 21-26.

Hillam, J. 1985b. Theoretical and applied dendrochronology: how to make a date with a tree. In: The Archaeologist and the Laboratory (ed Phillips, P.) CBA Research Report No. 58, pp. 17-23.

Hillam, J. 1987. (this volume) Problems of Dating and Interpreting Results from Archaeological Timber. pp. 141-155.

Hillam, J., Morgan, R.A. & Tyers, I.G. 1984. Dendrochronology and Roman London. Transactions of the London & Middlesex Archaeological Society 35 forthcoming.

Hollstein, E. 1980. Mitteleuropäische Eichenchronologie. Mainz: Zabern, pp. 273.

Huber, B. & Giertz, V. 1970. Central European dendrochronology for the Middle Ages. In: Scientific Methods in Medieval Archaeology (ed Berger, R.) London: University of California Press, pp. 201-212.

Hughes, M.K., Milsom, S.J. & Leggett, P.A. 1981. Sapwood estimates in the interpretation of tree-ring dates. Journal of Archaeological Science 8, pp. 381-390.

Morgan, R.A. 1982. Tree-ring studies in the Somerset Levels: The Meare Heath track 1974-1980. Somerset Levels Papers 8, pp. 39-45.

Morgan, R.A. 1984. Tree-ring studies in the Somerset Levels: the Sweet track 1979-1982. Somerset Levels Papers 10, pp. 46-64.

Orcel, C. & Egger, H. 1982. Analyse dendrochronologique des bois de la station littorale d'Auvernier-Port. Cahiers d'Archéologie 25, pp. 117-129.

Pilcher, J.R., Baillie, M.G.L., Schmidt, B. & Becker, B. 1984. A 7,272 year tree-ring chronology for western Europe. Nature 312, pp. 150-152.

Siebenlist-Kerner, V. 1983. Dating of architectural objects. In: Dendrochronology and Archaeology in Europe. (eds Eckstein, D., Wrobel, S. & Aniol, R.) Mitteilungen Bundesforschung. Forst- und Holzwirtschaft 141, pp. 195-208.

A Review of the Methodology for Calibrating Radiocarbon Dates
into Historical Ages

T.C. Aitchison and E.M. Scott

Department of Statistics
University of Glasgow
Glasgow

ABSTRACT

This review article describes the problems involved in the calibration of a radiocarbon date (in conjunction with its associated error) to a calendar age.

Recent investigations of the commonly quoted error of radiocarbon dates are discussed with their subsequent implications for calibration and currently available calibration schemes are critically compared.

Results suggest that the incorporation of realistic error terms in any calibration scheme produces a considerable increase in the width of the typical calibration interval. Even when using a high precision scheme to calibrate normal precision dates, there is only a minimal narrowing of such intervals.

A reduction in the width of the calibration intervals may be made by either calibrating a summary date obtained from replicate samples or, where possible, by calibrating a floating chronology.

INTRODUCTION

Although the use of radiocarbon dating in archaeology has increased considerably in the last decade, problems in the use and interpretation of radiocarbon dates have continued to arise. At the same time, the development of "high precision" laboratories has led to as yet unanswered questions about the "optimal" procedure for archaeologists wishing to make use of radiocarbon dating. In this review article we will consider the problems of calibrating radiocarbon dates into historical ages in the light of recent experimental and theoretical work.

The need to calibrate radiocarbon dates result from past fluctuations of atmospheric radiocarbon concentrations as measured in known-age tree ring samples (Suess, 1978; Stuiver, 1982; Baillie, 1983).

A calibration scheme is composed of three basic elements:

1: a series of samples of known true age with corresponding measured radiocarbon dates;

2: a procedure for estimating the form of the dependence of a radiocarbon date on its true age;

3: a method for calibrating the true age from a radiocarbon date.

We will consider these three elements in turn throughout the paper.

The primary source of data for calibration purposes has in the past been the long Bristlecone Pine series (Suess, 1978) but more recently several shorter "high precision" data sets have been published (Pearson et al., 1977; Bruns et al., 1980; Stuiver, 1982; Pearson et al., 1983) based primarily on European oak. The European dendrochronological records have recently been linked together to provide a continuous chronology of more than 7,000 years (Pilcher et al., 1984) and complete radiocarbon analysis of this sequence will ultimately provide an extensive matched radiocarbon and dendrochronological record for calibration purposes. It is hoped to publish this 7000 year series in a forthcoming special issue of Radiocarbon.

There are many alternative schemes for calibration of single dates and floating series. These differ in various important ways and part of this paper critically compares these methods with some resulting proposals for the archaeologist.

THE SIGNIFICANCE OF COMMONLY USED QUOTED ERRORS

The first problem is the significance of the quoted error. Of considerable interest and importance in its own right, this also exerts a strong influence in the area of calibration because resolution of this will, in part, make the age calibration more realistic. The commonly quoted errors contribute to calibrated age ranges in two ways; first by influencing the uncertainty on the fitted calibration curve itself and, secondly, by determining the uncertainty on the date to be calibrated. This section briefly presents some experimental results on the interpretation of the error associated with a radiocarbon date.

Interpretation of the experimental errors on radiocarbon age measurements for normal precision laboratories

It has long been acknowledged, though not always fully acted upon, that radiocarbon dating measurements are not definitive, i.e. they do not produce precise age estimates. Every radiocarbon age has an associated error term of which a major component is the counting error of the radioactive measurement process. Since the true age of a single sample is seldom known, the quoted counting error (i.e. one standard deviation) should be interpreted as meaning that over a long series of repeated counts of this sample, about 95% of the intervals of the form "measured radiocarbon age +/- twice the quoted error" should contain the true, but unknown, radiocarbon age of the sample. In routine dating practice, the quoted error terms are estimated in different ways by different laboratories but in most cases the dominant contribution derives from the counting error calculated from the number of random radiocarbon decay events measured during analysis. Many laboratories have devised error calculations which include components from various additional sources (Otlet, 1979; Pearson, 1979) although little experimental verification of such errors has been reported. There are however no agreed criteria among radiocarbon laboratories as to what components of variation should be included in the overall error to be quoted with any radiocarbon date.

Experimental results

In recent years it has become common for radiocarbon dates from several laboratories to be used in archaeological studies. This procedure may cause additional difficulties due to the possibility of inter-laboratory bias and differing precisions. Thus it is important that designed collaborative studies should become a regular and important part of laboratory procedure. Several such studies have already taken place and we describe their results briefly.

Firstly with regard to normal precision laboratories, a number have tried to assess the meaning of the commonly quoted errors in radiocarbon dating. Although their conclusions differ in specific detail, most agree that quoted errors based largely on counting errors considerably underestimate the true errors. For example, Clark (1975) investigated data obtained from several laboratories on Bristlecone Pine samples of similar ages and concluded that the true errors should be approximately twice the quoted values. Pardi and Marcus (1977) assessed similar measurements on archaeological samples and observed that, in some instances, true errors might be as much as four times those quoted. However, these two reports were based on results produced by many laboratories in a coincidental or fortuitous manner.

Nevertheless, results from _designed_ studies of both inter- and intra-laboratory nature have also confirmed the nature of these findings. Campbell and Baxter (1979) and Scott et al. (1983) found in the replicate analysis of homogenised wood samples that the commonly quoted error is frequently inadequate as a measure of true precision. Thus it is important for the archaeologist to ascertain from his local laboratory the exact derivation and interpretation of the error which has been quoted.

An initial intercalibration programme was instigated jointly by the British Museum and Harwell laboratories (Otlet et al, 1980), in which replicate counting samples of pure benzene of widely different apparent radiocarbon ages were assayed at eight UK laboratories. The results suggested that, for the radiocarbon counting procedure alone, quoted counting errors did adequately describe measurement variability. A second study, based in Glasgow (I.S.G., 1982 & 1983) extended replicate analysis to the entire dating process including pretreatment, sample conversion to the appropriate counting gas or liquid and radiocarbon counting itself. Analysis of the results from twenty laboratories throughout the world suggested that commonly quoted counting errors should be approximately doubled and that several of the laboratories which participated in the study were systematically biased with respect to others and to the overall trend by an amount up to several hundred years.

Secondly, with regard to high precision laboratories which have developed in recent years, these may quote errors of less than 20 years as opposed to the commonly quoted error of between 40 and 80 years (Brun et al., 1980; Pearson, 1980; Stuiver, 1982). These laboratories (e.g. Belfast, Seattle and Heidelberg) have been collaborating in providing a "high precision" calibration data set. The results so far published show an impressively close agreement. However, in one of the above mentioned studies (I.S.G.) systematic discrepancies of the order of 20 years between the high precision laboratories were found. Further, there was some indication that these laboratories also consistently underestimated their errors.

The development of high precision counting means that by 1986 there will exist a choice of two data sets for calibration, one where the errors on the dates are typically of the order of 50-70 years and a second with errors of 15-30 years.

The obvious questions which will occur to archaeologists are "what advantages will using dates from a high precision laboratory provide" and "how might one use dates from both a normal and high precision laboratory at a single site?"

First it must be said that many archaeological samples will prove unsuitable for high precision dating since typically the laboratory will require considerable quantities of original sample material. As we will see later, calibration of high precision dates on a high precision calibration scheme will bring considerable benefit in terms of reducing the "error" in the estimate of historical age. However, calibrating normal precision dates using a high precision calibration scheme will provide little extra benefit. For normal precision dates, the only means of "improving" on the errors in single dates is the use of either replicate samples of the same historical age or a floating sequence of sample material.

CALIBRATION OF A SINGLE DATE

The importance of the previous discussion of the quoted error lies largely in the critical role that these errors play in an appropriate calibration scheme. We must ask the following questions:

1: how have the errors on the radiocarbon dates from which the calibration scheme is derived been translated into corresponding uncertainties in the fitted calibration curve?

2: how are the errors on calibrated dates evaluated, including error contributions from the observed radiocarbon dates to be calibrated and from the fitted calibration function itself?

To answer these questions regardless of the data source used, we must also consider the following:

(a) the estimation of the central calibration curve based on either of the chosen data sources;

(b) the provision of suitable and realistic error bands on the estimated calibration curve;

(c) the error on the radiocarbon date to be calibrated.

Note that realistic errors on the radiocarbon dates in the chosen data source should be used in the construction of both (a) and (b) above.

Fitting the calibration curve

Initially the problem of calibrating radiocarbon dates to true ages involves the estimation of the form of dependence of a radiocarbon date on its true age. To achieve this we require

1: a data source of a series of samples of known true ages and corresponding radiocarbon dates

2: a formal method of modelling the form of dependence of radiocarbon date on true age.

The method used in (2) must always involve some element of subjective choice in terms of the assumptions of the underlying model. There are two basic areas of curve fitting techniques that could be used here, namely

(a) parametric procedures which specify explicitly the form of dependence, such as fitting polynomials (Damon et al, 1974; Klein et al, 1980) and

(b) non-parametric procedures such as Fourier smoothing (Houtermans, 1971; Klein et al, 1982), convoluted smoothing (Clark, 1975) and spline smoothing (Neftel, 1980).

The methods in (b) make little or no a priori assumptions about the form of dependence and are thus more flexible in allowing the data to determine the basic shape of the relationship. Such calibration systems may therefore be preferable although there has to be some choice of the degree of 'smoothing' performed, based on some general guidelines and a certain amount of subjectivity. The central calibration curve, fitted from the data from the master chronology, is thus only an estimate of the underlying curve and so has an error associated with it. As a result we must draw bands around the estimated central curve which quantify and describe our uncertainty about its shape.

The calibration procedure is intended, however, to make a quantitative statement, preferably in the form of a range of plausible values, about the true age of a sample given the observed radiocarbon date. Thus somehow we must reverse the inferential process used above in order to describe the uncertainty in radiocarbon age at a given true age. The essential information required for such a procedure is the estimated form of the central calibration curve and the width of the associated 'uncertainty bands'. The major differences between the recently proposed calibration schemes are really related to how these uncertainty bands are combined with the error on the date to be calibrated to provide, via the calibration, an error on the estimate of the true age of the sample.

Methods of calibration

Suppose we have a fitted calibration curve $x = \hat{f}(t)$ where x is the expected (i.e. long-run average) radiocarbon date at a true age of t, and $\hat{f}$ the estimated function describing the relationship between these two. Further assume that the error in the estimate of the expected radiocarbon date, x, at a true age of t is given by $s_f(t)$ i.e. the error in the fitted curve may vary with the true age.

Now let us attempt to calibrate a sample with measured radiocarbon date of x_c and corresponding true error s_c in an attempt to estimate the true, but unknown, age, u, of the sample.

Clearly a sensible point estimate, $\hat{u}$, of u is given by intersecting

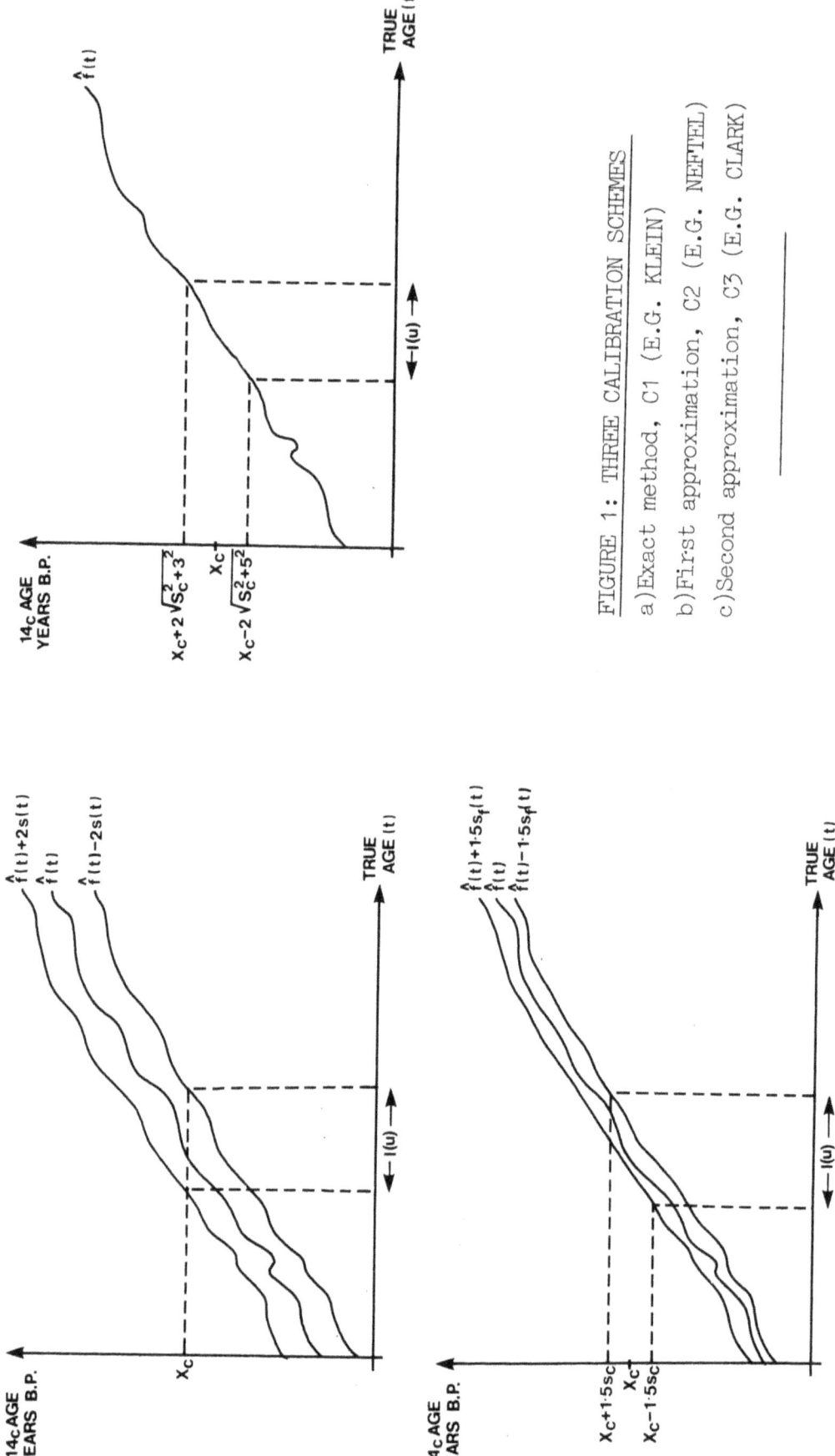

FIGURE 1: THREE CALIBRATION SCHEMES
a) Exact method, C1 (E.G. KLEIN)
b) First approximation, C2 (E.G. NEFTEL)
c) Second approximation, C3 (E.G. CLARK)

the line $x = x_c$ with the function $\hat{f}$ and then projecting down onto the horizontal axis (i.e. by effectively solving the equation $x_c = \hat{f}(u)$ for u). To get some idea of the uncertainty in $\hat{u}$ we could use one of the following three methods.

In the first method, C1, recently adopted by Klein et al. 1982, the calibration interval $I(u)$ is found from the intersections of the upper and lower curves, $\hat{f}(t) + 2s(t)$ and $\hat{f}(t) - 2s(t)$ where

$$s(t) = \sqrt{s_f^2(t) + s_c^2}$$

with the line $x = x_c$ which are then projected down onto the horizontal axis. Figure 1(a) indicates how this would be done. The major difficulty of such a procedure is that both the upper and lower calibration curves are different for different errors associated with the radiocarbon date. Thus in a sense new sets of calibration curves and charts would be required for each and every date to be calibrated, a technical requirement rendering the method unattractive for routine archaeological purposes.

The second method, C2 overcomes this problem, by approximating the calibration interval simply by using the 'uncertainty bands' on the fitted calibration curve and their intersection with the two lines of the form

$$x = x_c \pm ks_c$$

i.e. we find the intersection of $x_c + ks_c$ with $\hat{f}(t) - ks_f(t)$ and $x_c - ks_c$ with $\hat{f}(t) + ks_f(t)$ as illustrated in Figure 1(b). The choice of $k = 1.5$ gives calibration intervals with approximate 95% confidence i.e. roughly equivalent to the intervals of method C1. The method was used by Neftel (1980) in a widely circulated report, but he used $k = 2$ which gave far too conservative intervals for 95% confidence.

The third method, C3, is also approximate, and involves extending the uncertainty limits around the radiocarbon error on the central curve leading only the central calibration curve as shown in Figure 1(c) i.e. we find the intersections of

$$x_c \pm 2\sqrt{s_c^2 + \bar{s}^2}$$

with $\hat{f}(t)$ where $\bar{s}$ is a subjectively chosen and assumed constant error for the fitted calibration curve. One obviously unsatisfactory element to this approach involves the fact that the uncertainty in the fitted curve itself is by no means constant across true age and it is therefore quite contentious as to what value of $\bar{s}$, the average error in the fitted curve, should be used for the purposes of any single calibration. This method was in fact adopted by Clark (1975), Stuiver (1982) and Pearson et al. (1985).

These approximate methods are particularly appealing in practice since they require provison of only one calibration chart (or set of tables) but the degree of approximation involved must clearly be examined with implications for the quoted level of confidence, and the width of the final calibrated true age range.

Existing calibration procedures compared

It is valuable to compare and contrast the nature of the error treatment of these various calibration schemes and to consider some of the various discussions which have taken place in the literature.

In general we should note that

(a) the result of the calibration will in all three methods be an interval for the true age, since both the radiocarbon date and the calibration curve are known to be subject to error.

(b) in sections where the calibration curve slopes steeply upwards, the calibrated interval will be narrow. Where the calibration curve is "flat" or slopes very gradually, the calibrated interval will be wide.

(c) since the calibration curve may be very jagged in appearance, multiple, non-overlapping calibrated intervals may on occasion be produced if the error on a date is small.

(d) examination of published calibration schemes shows immediately that they differ first of all in the actual method used to estimate the central calibration curve. Thus differences in the resulting calibration intervals will be due not only to the use of different methods but also to the method of curve fitting employed.

More particularly, the schemes also differ markedly in their use of the reported errors on the radiocarbon dates of known true ages. Clark's scheme actually ignores the quoted errors, preferring to use estimated "blanket" errors which are dependent not on the quoted error but on the radiocarbon age. In contrast, Klein et al.'s scheme adjusts the quoted error to a claimed "more realistic" level based on a slowly increasing function of sample age (although no experimental justification of this function is given). Stuiver's high precision calibration system is similar in nature to Clark's, i.e. it incorporates the error on the date to be calibrated, but the author gives no details of the method of estimating the calibration function. He does, however, quote an associated error on the calibration curve of +/- 6 years regardless of true age. This calibration scheme is based on a 2000 year chronology and so is shorter than the others which are derived from the entire Bristlecone Pine record, but it is included here since it is presumably representative of future high precision chronologies.

We feel that the choice between these published calibration schemes is difficult to make but should be based on the ease of use of the calibration method provided. The method adopted by Klein et al. is likely to achieve some form of acceptance amongst radiocarbon analysts but is unlikely to attract the favour of radiocarbon users, particularly in its use of bands for the curve which are dependent on the quoted error of the sample date to be calibrated. The method adopted by Clark is, in comparison, much simpler to use but has the disadvantage of not using the sample quoted errors in the construction of thecurve and in the actual calibration procedure. On the other hand, a calibration procedure similar to that adopted by Neftel answers both these criticisms while the procedure remains easy to use. This procedure may be used for any calibration curve where confidence bands for the curve are provided.

The level of confidence to be associated with calibration intervals obtained from any of these methods cannot be completely justified primarily due to the choice of procedures in the 'smoothing' of the central calibration curve. However, for most true ages this confidence should not be too far from 95% for any of these schemes and, in particular those based on method C1 for calibration, are at least arguably using the theoretically correct approach to obtain 95% confidence. However, we must stress that, regardless of data set or calibration curve estimation employed, a user is still faced with the choice of calibration method as illustrated in Figure 1.

A comparison of calibration intervals obtained using the schemes of Klein, Neftel, Clark and Stuiver was reported in Scott et al. (1984). Various radiocarbon dates with differing errors were calibrated using the different approaches and the resulting calibrated intervals compared. It should be noted however that the differences in the calibration intervals for these four methods reflect not only the different methods of calibration used but also the fact that they are all based on different techniques for fitting the calibration curve as well as being applied to different master chronologies.

The conclusions of the study were

1: Differences in resulting intervals do exist over all values of radiocarbon age and, in general, the widths of the calibrated intervals (on average about 600 years with a range of 100-800 years) are wider than would presumably be hoped for by the archaeologist.

2: Using a high precision scheme for calibrating normal precision dates is of little advantage since the widths of the calibrated intervals range from 100-430 years. The intervals obtained when calibrating high precision dates are certainly fairly narrow (typically - 50 years) but often the calibrated age range is in the form of disjoint intervals.

3: Intervals produced by Klein's scheme are generally narrower than those from the schemes of Neftel and Clark. For larger quoted errors, Neftel's scheme produces very wide intervals which is not surprising given the ultra conservative choice of $k = 2$ for Neftel's scheme.

4: Although, in general, the widths of calibrated intervals tend to increase with sample age, the specific position on the curve does influence the final range considerably (e.g. the intervals for a 1000 BC sample are consistently wider than those for a 3000 BC sample).

5: It is however evident that, in general, and even using conventionally quoted error terms on the radiocarbon dates, the calibrated ages should have considerable uncertainties associated with them. Thus in terms of archaeological relevance the differences between the schemes are in most cases small in comparison to the actual widths of the calibrated intervals.

6: Inclusion of more realistic errors, using results from the various experimental studies resulted in larger calibrated intervals and even here the use of high precision calibration schemes for normal precision dates appeared to be of little advantage.

In summary then for calibration of single dates, it is clear that the usefulness of single radiocarbon dates is severely curtailed.

An additional difficulty with any of the above mentioned calibration schemes concerns calibration of results from samples with quite different growth periods from the master chronology data. Where this type of information is available, some suggestions have been made by Mook et al. (1979) and de Jong (1980). They adopt an approach which involves taking the fitted calibration curve (perhaps based on single ring results) and smoothing it by moving average procedures, where the degree of smoothing depends on the years of growth of the sample to be calibrated. Fortunately calibration into true ages of a less precise analysis requires less certainty about the growth period represented by the sample. The above authors conclude that a calibration curve should be used which applies to the number of years represented by the sample, and this becomes essential the more precise the radiocarbon measurement.

Implementation of "high precision" calibration chronologies (e.g. Stuiver's) is found to produce no substantial reduction in these intervals for normal precision dates. For calibration of "high precision" dates however, such chronologies are of considerable value. Thus, given an extensive "high precision" calibration scheme is currently under production, there are considerable benefits to be derived from parallel progress towards reducing errors in "normal precision" dating laboratories. If the latter trend does not occur, the archaeologist will gain little from the high precision collaborative work.

CALIBRATION WITH IMPROVED PRECISION: REPLICATE SAMPLES AND FLOATING CHRONOLOGIES

We now consider the calibration of

1) a single summary radiocarbon date based on replicate samples

and (2) a floating chronology of samples of known differentials in true ages but an unknown 'absolute' age.

Replicate Samples

If we have sufficient sample material to divide into several sub-samples or a collection of samples known to have effectively the same true age (e.g. within a span of 50 years) then a substantial increase in the precision of the calibrated age can be obtained by dating these sub-samples separately.

Ward and Wilson (1981) describe a very simple method of combining the radiocarbon dates of such replicate samples into a summary 'radiocarbon date' taking into account the appropriate errors in all the dates. This summary data is in fact just a weighted average of the radiocarbon dates of the replicate samples and its associated error will be considerably less than that of any of the samples themselves. These authors also provide a formal hypothesis test of whether or not the replicate samples could indeed be assumed to have the same underlying radiocarbon, and hence, true age.

In terms of calibration we are now simply back in the position of having a single, albeit summary, radiocarbon date with its associated error and hence we can calibrate by exactly the same scheme chosen from the previous section.

Floating Chronologies

The second method of gaining increased precision in the calibrated dates is that of calibrating a floating chronology i.e. the placement of a chronology in "absolute" historical age by finding an interval estimate for, say, the youngest part of the floating chronology which we will denote by α. The advantages of calibrating a floating chronology have long been recognised. For example, Ferguson et al (1966) describe a procedure for matching the time derivatives of two chronologies. However these authors used a subjectively drawn curve and provided no methods for estimating the precision of their estimate of α. Their basic ideas can nevertheless be formalised and extended as follows.

The ingredients for calibrating such a floating series of dates are similar to that for calibrating a single date, namely the fitted calibration curve from a master chronology and its appropriate uncertainty bands. Also, since the series and the fitted calibration curve have associated uncertainties, we can only hope to produce a range of plausible values for α.

Clark and Renfrew (1972) described a method which assumes a locally linear relationship between the radiocarbon and true ages, with the additional assumption that the floating and master chronologies have the same error for each radiocarbon date. This assumption clearly would be invalid for calibrating a high precision series against the Bristlecone Pine chronology.

Their simple procedure involves fitting two parallel straight lines and manipulating the resulting parameter estimates to give an estimate of α. More importantly, this approach can provide a confidence interval for α. An important extension of this rather basic method allows for calibration of two floating series which is clearly of some importance when there are known relationships between the series e.g. one is 200 years older than the other.

In a later paper (Clark and Sowray (1973)) the assumption of linearity was replaced by the less rigid assumption that over the short period considered, the forms of the master and floating chronologies should be the same smooth curve. They give details of this approach in the case of fitting piecewise polynomials but as usual the shape of the curve is not determined by the data alone.

Kruse et al. (1980) described a further method for calibrating a floating chronology, which differs in the method of curve fitting from that suggested by Clark, and differs in the way an estimated value for α is found. A smooth cubic spline function (a non-parametric method of curve fitting) is fitted through the master data or through a section thereof. The estimate of α is that value which in a least squares sense best fits the floating chronology to the fitted calibration curve. The authors comment that it would be difficult to derive an explicit expression for the error on the estimate of α and use only this point estimate in any applications which they describe. This is clearly unsatisfactory.

The final two methods are important since they provide a means for assessing the error on the estimate of α. The approaches described by Scott et al. (1981) and Clark and Morgan (1983) provide interval estimates for α based on the uncertainty on the calibration curve and on the floating series. Roughly speaking both approaches provide a range of values for α made up of the set of values for α for which the scatter of the floating chronology, when fixed at that value of α on the fitted calibration curve, is in agreement with the quoted errors on the floating chronology and the uncertainty on the calibration curve.

CONCLUSIONS

In summary then, we feel the following four points should be considered by the archaeologist when planning to calibrate radiocarbon dates and interpret their results.

(1) Realistic errors only should be used and these are likely to be at least twice the counting error. Thus radiocarbon laboratories should be encouraged to participate in cross checking to remove any bias, and to specify how their quoted errors are derived.

(2) Confidence bands on the fitted calibration curve provide a simple and adequate basis for calibration of radiocarbon dates and provision of appropriate errors on the calibrated ages.

(3) Calibration intervals so obtained will generally be very wide with the exception of calibration of high precision dates on a high precision chronology.

(4) Use of replicate samples and floating chronologies are the only means to improve the 'precision' of calibrated dates.

REFERENCES

Baillie, M.G.L. 1983. Dendrochonology: the current situation. (In B.S. Ottoway, ed.) *Archaeology, Dendrochronology and the Radiocarbon Calibration Curve.* University of Edinburgh Press, pp. 15-25.

Bruns, M., Munnich, K.O., Becker, B 1980. Natural radiocarbon variations from 200-800 AD. *Radiocarbon* 22(2), pp. 273-278.

Campbell, J.A., Baxter, M.S. 1979. Radiocarbon measurements on submerged forest floating chronologies. *Nature,* 278, pp. 409-413.

Clark, R.M. 1975. A calibration curve for radiocarbon dates. *Antiquity,* 49, pp. 251-266.

Clark, R.M., Renfrew, C. 1972. A statistical approach to the calibration of floating tree-ring chronologies. *Archaeometry* 14, pp. 5-19.

Clark, R.M., Sowray, A. 1973. Further statistical methods for the calibration of floating tree-ring chronologies. *Archaeometry* 15, pp. 255-266.

Clark, R.M., Morgan, R.A. 1983. An alternative statistical approach to the calibration of floating tree-ring chronologies: two sequences from the Somerset levels. *Archaeometry* 25, pp. 3-16.

De Jong, A.F.M. 1980. *Natural radiocarbon variations.* PhD Thesis, Univ. of Groningen, The Netherlands.

Damon, P.E., Ferguson, C.W., Long, A., Wallick, E.I. 1974. Dendrochronologic calibration of the radiocarbon timescale. *American Antiquity* 39, pp. 350-366.

Ferguson, C.W., Huber, B., Suess, H.E. 1966. Determination of the age of Swiss lake dwellings as an example of dendrochronologically calibrated radiocarbon dating. *Zeitschrift fur Naturforschung* 219, pp. 1173-1177.

Houtermans, J.C. 1971. *Geophysical interpretations of Bristlecone Pine radiocarbon measurements using a method of Fourier analysis for unequally spaced data.* PhD Thesis, Univ. Bern, Switzerland.

International Study Group (I.S.G.) 1982. An inter-laboratory comparison of radiocarbon measurements in tree rings. *Nature* 298, pp. 619-623.

I.S.G. 1983. An international tree ring replicate study. In: *Radiocarbon and Archaeology,* PACT 8, pp. 123-133.

Klein, J., Lerman, J.C., Damon, P.E. Linick, T.W. 1980. Radiocarbon concentrations in the atmosphere: 8000 year record of variations in tree rings. First results of a USA workshop. *Radiocarbon* 22(3), pp. 950-961.

Klein, J., Lerman, J.C., Damon, P.E., Ralph, E.K. 1982. Calibration of radiocarbon dates: Tables based on the concensus data of the Workshop on calibration of the radiocarbon timescale. *Radiocarbon* 24(2), pp. 103-151.

Kruse, H.H., Linick, T.W., Suess, H.E., Becker, B. 1980. Computer matched radiocarbon dates of floating tree ring series. Radiocarbon 22(2), pp. 260-267.

Mook, W.G., De Jong, A.F.M., Geertsema, H. 1979. Archaeological implications of natural radiocarbon variations. Palaeohistoria 21, pp. 9—18.

Neftel, A. 1980. The construction of a radiocarbon calibration curve based on radiocarbon measurements of absolutely dated American tree ring samples. Unpublished report, Univ. Bern, Switzerland.

Otlet, R.L. 1979. An assessment of laboratory errors in liquid scintillation methods of radiocarbon dating. In: Berne, R. Suess, H.E. (ed.) Proceedings of the 9th International Radiocarbon Conference Univ. California Press.

Otlet, R.L. Walker, A.J., Hewson, A.D., Burleigh, R. (1980). Radiocarbon interlaboratory comparisons in the UK: Experiment design, preparation and preliminary results. Radiocarbon 22(3), pp. 936-946.

Pardi, R., Marcus, L. 1977. Non-counting errors in radiocarbon dating. Annals New York Academy of Sciences 288, pp. 174-180.

Pearson, G.W. 1979. Precise radiocarbon measurement by liquid scintillation counting. Radiocarbon 21, pp. 1-21.

Pearson, G.W. 1980. High precision radiocarbon dating by liquid scintillation applied to radiocarbon timescale calibration. Radiocarbon 22(2), pp. 337-349.

Pearson, G.W., Pilcher, J.R., Baillie, M.G.L., Hillam, J. 1977. Absolute radiocarbon dating using a low altitude European tree ring calibration. Nature 270, pp. 25-28.

Pearson, G.W., Pilcher, J.R., Baillie, M.G.L. 1983. High precision radiocarbon measurement of Irish oaks to show natural radiocarbon variations from 200BC to 4000BC. Radiocarbon 25(2), pp. 179-186.

Pilcher, J.R., Baillie, M.G.L., Schmidt, B., Becker, B. 1984. A 7,272 year tree ring chronology for western Europe. Nature 312, pp. 150-152.

Scott, E.M., Baxter, M.S., Aitchison, T.C. 1981. An assessment of variability in radiocarbon dating. In: Methods of Low Counting and Spectrometry. IAEA, Vienna, pp. 371-392.

Scott, E.M. Baxter, M.S., Aitchison, T.C. 1983. Radiocarbon dating reproducibility: Evidence from a combined experimental and statistical programme. In: Radiocarbon and Archaeology, PACT 8, pp. 133-147.

Scott, E.M., Baxter, M.S., Aitchison, T.C. 1984. A comparison of the treatment of errors in radiocarbon dating calibration methods. Journal of Archaeological Science, 11, pp. 455-466.

Stuiver, M. 1982. A high precision calibration of the AD timescale. Radiocarbon 24(1), pp. 1-27.

Suess, H.E. 1978. La Jolla measurements of radiocarbon in tree-ring dated wood. Radiocarbon 20, pp. 1-18.

Ward, G.K., Wilson, S.R. 1981. Evaluation and clustering of radiocarbon age determinations: Procedures and paradigms. Archaeometry 23(1), pp. 19-39.

The Belfast 'Long Chronology' Project

M.G.L. Baillie and J.R. Pilcher

Palaeoecology Centre
Queen's University
Belfast
Northern Ireland

ABSTRACT

An oak tree-ring chronology of over 7000 years has been completed using timbers from the British Isles. Nearly 1000 timbers, mostly from the north of Ireland, were combined into the chronology which for most of its length has a replication of between 10 and 20 trees. The chronology forms a standard for calibration of the radiocarbon timescale.

INTRODUCTION

This article aims to document, in an historical fashion, the construction of a long oak tree-ring chronology in Ireland. It provides additional historical background and detail to the short account previously presented (Pilcher et al 1984).

In 1968 one of the authors (JRP) in collaboration with the then director of the Palaeoecology Laboratory, Dr. A.G. Smith, decided to investigate the possibility of constructing a long Irish tree-ring chronology to provide a European standard for calibration of the radiocarbon timescale. Two principal factors brought about this decision. One was the abundance of sub-fossil timbers in the north of Ireland recently brought to light by motorway construction and land reclamation. The other was the availability of a radiocarbon dating facility within the Palaeoecology Laboratory. The stimulus lay in the radiocarbon calibration produced by Suess (1965, 1970) which had been based on bristlecone pine wood from California (Ferguson, 1968, 1969) and had left two main questions unanswered. One was whether the calibration based on high altitude trees from California could be applicable to Europe. There was a strong body of archaeological opinion that was loath to accept the revision of their timescales which the calibration suggested. The second problem related to the short-term fluctuation or 'wiggles' in the Suess calibration which gave rise to an extensive literature during the 1970's (e.g. Renfrew and Clark, 1974; Clark, 1975; McKerell, 1975). If the wiggles were real then the value of radiocarbon dating as an archaeological tool was considerably diminished. There was a strong instinctive reaction against the evidence produced by Suess. Thus it was clear that an 'Old World' calibration curve was needed.

As the first step towards this aim one of the authors (MB) was employed from mid-1968 to look into the feasibility of building a tree-ring scale for oak in Ireland. At the same time G. W. Pearson came to the Palaeoecology Laboratory to run the methane radiocarbon dating system. By 1970 a series of investigations on living trees and historic building timbers had demonstrated that oak tree-ring patterns could be reliably cross-dated in the north of Ireland (Baillie, 1973) and methods and practices applicable to the dendrochronology of oak were established. On the strength of these findings the project was divided: one 'half', based on archaeological and historic building timbers (Baillie, 1974, 1975, 1977a, 1977b), attempted to build a chronology back from the present covering the last two millennia, whilst the other was aimed at the BC era and was based almost exclusively on sub-fossil oaks from bogs, lakes and river beds.

TOOLS FOR THE JOB

Tree-ring research depends on the ability to identify correctly the place where two separate ring patterns coincide. Human subjective assessment requires independent verification and attention was paid to statistical tests that would perform this function. These tests were built into a series of computer programs developed at Hamburg (Eckstein and Bauch, 1969) and Belfast (Baillie and Pilcher, 1973). The Belfast program (CROS) took account of the magnitude of the year-to-year variations in the ring patterns and, although it was not perfect in a statistical sense, it operated successfully in picking out unique matching positions for long overlaps (for more information on short overlaps and CROS limitations see Pilcher and Baillie this vol). The program was always used in its original form throughout the project and this allowed all correlation values to be compared as they were all calculated by a standard statistic i.e. there was no question of tuning the procedure to allow for special cases - all timbers had to be treated alike.

Failure of the human observer and CROS to pick out the same overlap position led to a mutual veto system where a match would not be accepted if either 1) a visual match was not supported by CROS or 2) a high computer correlation was not visually acceptable to the dendrochronologist. These harsh criteria were particularly important in dealing with the prehistoric bog oaks simply because these chronologies were going to be difficult to check externally. The matching procedures in combination with between-site replication had to be sufficiently robust to produce a system which could be checked internally. Generally, chronology building was restricted to timbers which showed extremely good cross-dating.

Having established a set of procedures for preparation, measurement and cross-dating, our success was dependent on the natural distribution of oaks through time. Why should timbers be preserved from all periods? It seemed likely that there would be some periods of abundance i.e. when bogs were capable of supporting oaks, and other periods from which little or no material would survive. In order to test for such clustering a series of essentially random samples was submitted for routine radiocarbon analysis. By the early 1970's evidence suggested that the bog oaks were spread reasonably uniformly through time (Smith et al, 1972). Nothing remained but to go ahead and attempt to build chronologies.

THE NORTH OF IRELAND SUB-FOSSIL CHRONOLOGIES

The basic story of the sub-fossil chronologies is contained in Pilcher et al. (1977). By working with groups of material from specific locations, a single small bog or area within a bog, it was found that long site chronologies, from 500 to 900 years in length, could be constructed with relative ease. The basic procedure was to compare each ring pattern as it was measured with all of the other ring patterns from the same site. This was continued until groups of timbers which clearly cross-matched - with long overlaps - were isolated and sub-masters were created. As more and more trees were studied, so the site chronology or chronologies were refined. Although this sounds straightforward, on some sites many individual trees had to be studied before even two were found to cross-date.

It is worth noting that in compiling the sub-fossil chronologies the bog oaks were put through a number of 'filters'. In the field any timber with less than about 100 rings was discarded as it was unlikely that it could be matched at all (by our procedures) and even if matched, was not going to extend any chronology. Secondly, failure of a timber to survive the 'mutual veto' led to the timber being discarded.

A third filtering procedure employed at Belfast came to light only by comparison with procedures in German Laboratories. Our policy has always been to measure one 'optimum' radius. Only if a timber had some obvious problem feature, or if a long ring pattern consistently failed to cross-date, was re-measurement normally considered. In contrast, a common German procedure was to measure several radii, the 'ring pattern' being the mean of these. At first sight this German procedure is 'better' in that the mean of several measurements is better than a single measurement. However it has the disadvantage that three times the amount of effort goes into the processing of each timber. Put another way - the Belfast procedure allows for the processing of three times the amount of material. It may be wasteful - because some material will not cross-date - but with the large number of bog oaks available we could afford to take this approach and select only the best matches. In essence this single measurement procedure was acting as a filter for the best ring patterns.

Once site chronologies had been constructed, the problems then centred on joining these site units together to form longer multi-site chronologies. This work proceeded steadily during the early 1970s but was pushed forward dramatically in the mid-70s with the discovery of the important sub-fossil site at Garry Bog in Co. Antrim. Extensive land-reclamation work in this area in 1975-6 yielded timbers from most periods from 5000 BC down to 200 BC. This was particularly important because previously almost no material had been obtained for the 1st millennium BC.

By 1976, chronologies covering the BC era consisted of:

a) Three chronologies of 600, 700 and 800 years, all pre-4000 BC.

b) The published "long chronology" covering 2990 years from approximately 4000 to 1000 BC (Pilcher et al. 1977) and

c) The "Garry Bog 2" chronology of 719 years spanning approximately 900 to 200 BC.

Thus within six years we had filled virtually the whole of the first five millenia BC. By this stage the chronology building had changed in character. In order to complete the chronology it was necessary to obtain timbers of specific ages. Searching for such timbers is more or less impossible since the timbers give no clue to their age in the field.

In theory random sampling would eventually have given us the timbers we required. However the cost effectiveness of such an approach which might have entailed many years of duplication of effort - with no guarantee of success - made this an unattractive proposition. In addition some evidence pointed to the possibility that the 900 BC gap was a real hiatus in the preservation of oak timbers at least in the north of Ireland. It appeared that the same gap occurred at no less than three different bog sites (Baillie, 1979b, 1982). Obviously such a finding had worrying consequences, as it was possible that the remaining gaps could not be bridged within the north of Ireland.

THE AD CHRONOLOGIES.

By the time the gaps in the prehistoric chronology had been defined, the Belfast absolute chronology had been constructed back to AD 1001 (Baillie, 1977a). A parallel section of Dublin chronology had been dated to AD 855-1306 (Baillie, 1977b). Earlier than this there was a 795 year chronology covering the approximate period AD 1-800 (Baillie, 1975). Thus the late 1970s saw most effort being put into the linking of these chronologies and the bridging of this 9th century AD gap. As the search was stepped up for archaeological material from the later 1st-millennium, it became clear that our horizons had to be lifted. We could no longer restrict our source area to the north of Ireland. As a response to the need to understand more about the areal extent of dendrochronological matching, localised site chronologies were constructed from living-tree cores from around Ireland (Pilcher and Baillie, 1980a). It became clear from this work that oaks within Ireland were responding to a common signal and that it was possible to demonstrate direct cross-dating between sites 400 km apart.

By late 1979 the collection of archaeological material from a series of horizontal mill sites from throughout Ireland had extended the northern floating chronology to 907 years and yielded a parallel southern chronology of 621 years, both ending in the later-9th century AD. It was becoming clear that it was going to be very difficult to bridge the 9th century gap with Irish material.

By this time it was known that good cross-dating could be observed between most Irish, English and Scottish chronologies in the 2nd millennium AD (Baillie, 1978). Again studies on a grid of modern oak chronologies from the whole of the British Isles (Pilcher and Baillie, 1980b), indicated a general response to some common signal (Baillie, 1982, 1983a). This suggested the possibility of utilizing chronologies from outside Ireland. With the acquisition of a precisely dated ring pattern from Tudor St. London (supplied by Jennifer Hillam), which spanned 682-918 AD, further efforts were made to tie down the 907 year floating chronology. This was accomplished in early 1980 with the processing of a timber from Ballydowane, Co. Waterford which matched both Tudor St. and the Irish chronology. This specified the dates of the northern chronology as AD 894 to 13 BC (Baillie, 1979b, 1980, 1981).

We were able to confirm the correctness of our chronologies by step-wise correlation from Ireland to England to the absolutely dated German chronologies. This had been observed in the case of the England/Wales border chronology (Siebenlist-Kerner, 1978), in the case of Fletcher's Ref 6 (Fletcher, 1977) and was now observed in the early mediaeval period for chronologies from Mersea (J. Hillam, pers. comm.) and Fletcher's Ref 8

(Fletcher, 1977, Baillie, 1980; Baillie, 1982; Baillie et al 1983). More recently the earliest part of the Irish AD chronology has shown good stepwise agreement to German chronologies via the London Southwark chronology (I. Tyers, pers. comm.). So by 1980 there was a continuous, independent, British Isles oak chronology completed and checked back to 13 BC.

THE SECOND PUSH ON THE LONG CHRONOLOGY

With the successful completion of the AD chronology emphasis shifted back to the problems of the prehistoric chronologies. From 1980 a major reworking of the three pre-4000 BC chronology sections together with the processing for the first time of a large collection of material from a site at Lough McGarry, Co. Antrim rapidly consolidated the three short floating sections into a single 1550 year floating chronology. By 1981 this chronology had been linked to the pre-existing 2990 year "long" chronology giving a total unit of 4341 years covering the approximate period 5300 to 1000 BC.

The situation had therefore simplified itself to two remaining gaps in more than seven millenia, which occurred between 13 BC and circa 200 BC and around 950 BC. As noted above it had begun to look as though this latter gap might reflect some real depletion event in the Irish bog oak record. The 13 BC to circa 200 BC gap simply reflected the total lack of sub-fossil material of this period coupled with the total absence of waterlogged archaeological sites of the early centuries AD. The only archaeological material in any way relevant to this period consisted of Iron Age timbers from the Navan and Dorsey sites in Co. Armagh which were believed to belong to the early centuries BC.

One possible approach was to attempt to date the long prehistoric chronology directly to established German oak chronologies. In 1980 Hollstein published a chronology back to 724 BC (Hollstein, 1980) and in 1981 Becker produced another more southerly German chronology back to 370 BC (Becker, 1981). It was not expected that cross-dating could be established from Ireland direct to Germany. The approach was to acquire any available batches of prehistoric material from Britain and attempt to use these as a step-wise link from Ireland to Germany.

In late 1980 bog oak samples were collected from the area around Midlenhall and Lakenheath in Suffolk. When processed these gave a 509 year chronology which matched directly with the 4341 year floating chronology and dated to the mid-third millennium BC. Unfortunately no direct dating could be achieved against available floating German chronologies.

The next significant batch of material was spotted from the window of the London to Durham train in March 1981. The site was located to Swan Carr farm near Sedgefield about ten miles south of Durham. It was visited in the summer of 1981 and the 20 initial samples gave a chronology of 775 years. Although our prime interest was in a link to Germany, the date being of secondary importance, this 755 year chronology dated against Garry Bog 2. More importantly it extended back in time across the Garry Bog 2/Long chronology gap and matched with the Long chronology. So as luck would have it our attempt to circumvent the 900-1000 BC gap, by linking our chronologies to Germany via England, led to the direct bridging of the gap with the Swan Carr chronology (Baillie et al 1983; Baillie, 1983b). The importance of this was that it kept the extended 5061 year chronology unit independent of the German work thus avoiding the likelihood of any circular argument. Coincidentally it was found that only one year separated the Long and Garry Bog 2 chronologies.

THE COMPLETION OF THE BELFAST LONG CHRONOLOGY

From 1980 we had been processing timbers from the Roman site at Carlisle in northern England in order to try bridging the only remaining gap at 13 BC to circa 200 BC. Long lived timbers from the 1st or 2nd centuries AD might solve the problem by running back across the gap. Carlisle turned out to be the most difficult group of timbers so far encountered at Belfast. The timbers behaved as if they had been drawn by the Romans from a variety of widely separated sources. However by 1982 a chronology had been constructed which appeared to date from AD 90 to 247 BC.

This dating relied on a link to the London Roman chronology from New Fresh Wharf which had been constructed by Ruth Morgan (pers. comm.) and in a sense it was unsatisfactory (from the point of view of independence) because New Fresh Wharf had been dated against German chronologies. This situation was later rectified when a composite Southwark chronology was put together by Ian Tyers which spanned AD 255 to 252 BC. The Southwark chronology dated directly with the older part of the Irish AD chronology and formed an effective British Isles link.

By 1982, therefore, we had an absolute chronology back to 247 BC and a 5061 year chronology which on radiocarbon evidence was likely to end around 200 BC. The only other relevant material was the Navan/Dorsey 246-year chronology which, again on radiocarbon evidence, should end in the first centuries BC. Interestingly there were reasonable visual matches with significant 't' values suggesting a link from Carlisle to Navan/Dorsey to the end of the Garry Bog 2. This tentative match suggested that Navan/Dorsey ended in 116 BC and Garry Bog 2 ended in 229 BC. If this was correct then the overall chronology would be complete back to 5289 BC.

The German Links

In 1981 and 1982 a series of German publications (Schmidt, 1981, Schmidt and Schwabedissen, 1982; Becker and Schmidt, 1982) indicated that the overall German chronology complex had been extended first to 1462 BC and then to 2061 BC. This had come about by the amalgamation of Becker's south central German chronologies with those from northern Germany constructed by Schmidt.

In the spring of 1982 it was agreed with both Bernd Becker and Burghart Schmidt that the time was right for the mutual exchange of data in an attempt to confirm the Irish and German chronologies. By the early summer of 1982 this exchange had begun and immediately Schmidt was able to demonstrate a long section of consistent cross-dating between the Irish and North German chronologies in the 2nd millennium BC. This agreement was sufficiently good to imply that our whole 5061 year prehistoric chronology ended in 158 BC. This direct link to the absolutely dated German chronology should have bypassed the problems in the first centuries BC, and in theory the arrival of Schmidt's telex with the news of that match should have marked the 'eureka' finish to the whole Long chronology programme.

Unfortunately, analysis of the available British Isles chronologies lent no support to the "158 BC" dating. This represented a serious dilemma in the Belfast Laboratory for the following reasons. All of our chronology building had been based on the assumption that when we had long overlaps between chronologies in our area (or indeed within most of the British Isles) then we would find visually and statistically significant cross-dating <u>at the correct relative positions</u>. If this German dating were correct then we had <u>overlaps</u>, between Garry Bog 2, Navan/Dorsey and Carlisle, but no significant visual or statistical cross-dating at the correct positions. If our procedure had broken down here how were we to know that it had not broken down elsewhere in our long chronology sections? So this 'test' of our procedure had profound implications.

However, our procedure had allowed the construction of a chronology back to 13 BC and then to 247 BC which relied on material from Ireland and England and which was independent of the German chronologies. It could be argued therefore that there was nothing wrong with our procedure, which tentatively indicated an end date for the 5061 year chronology at 229 BC. This dilemma forced us to examine the possibility that there might be something amiss with the German dating. This situation was forced to a head in the weeks leading up to the Seattle Radiocarbon conference in June 1982. Gordon Pearson's high precision calibration results, derived from sections of the 5061-year floating chronology, were to be presented at the Conference. As the diagrams were being drawn up the dendrochronologists were asked where the tree-ring axis should be placed. Should it be given a floating scale or should dates be specified? As a compromise the scale was numbered in 'Dendroyears BC' and the data was placed consistent with a 229 BC end year for Garry Bog 2 (Pearson <u>et al</u> 1983).

The re-examination of the whole dating problem caused by this exercise suggested to us that the 229 BC dating virtually had to be correct. However the German workers saw no reason to doubt their chronologies. An obvious dilemma was developing. Was the German prehistoric chronology complex correct or was the tentative Belfast chronology correct? This caused a detailed look to be taken at the German chronology. Clearly there was no doubt about the German chronologies back to around 400 BC as they were replicated by different workers (Hollstein, 1980; Becker, 1981). Also both the German and Irish workers were agreed that there was good agreement between the Irish and German chronologies in the 2nd millennium BC. Any doubt therefore had to relate to the period in between. A detailed look at the available German information indicated that their weakest point was around 550 BC where neither Becker nor Schmidt had continuous chronologies and where the whole German complex relied on a single site, Kirnsulzbach, originally published by Hollstein in 1973. Sufficient evidence existed to allow the postulation that there was an error of 71 years in Hollstein's chronology at that point and that the German chronology complex should be broken at 550 BC and all of their chronologies before that date should move back in time by 71 years (Baillie, 1983b). Hollstein could have fallen foul of "the dendrochronologist's dilemma" (Baillie <u>et al.</u> 1983) in trying to link his Kirnsulzbach chronology onto the end of a pre-existing chronology when no significant overlap existed. In 1984 this re-dating of the German chronologies earlier than 550 BC was agreed by Schmidt and Becker and a consensus reached on the existence of a continuous chronology of 7272 years (Pilcher <u>et al.</u> 1984).

THE RADIOCARBON CALIBRATION

The original intention had been to perform the radiocarbon calibration on the existing methane equipment at Belfast. We now know that such a re-calibration would have been a disastrous waste of time. The levels of precision and accuracy of which such routine equipment was capable would almost certainly have left the Belfast results open to the same criticisms as those of Suess. Thus, as the chronologies progressed in the early 1970s, Gordon Pearson set out to investigate the concept of "absolute" or "high-precision" radiocarbon measurement by liquid scintillation counting (Pearson, 1979, 1980). This work aimed at the measurement of radiocarbon ages with realistic precision of $\pm$ 20 years. It involved the assessment of all the variables likely to cause inaccuracy within the radiocarbon activity measurements.

Because the calibration could not be held off until the tree-ring chronologies were complete, samples were initially supplied from the central millennium of the 2990-year floating chronology. The samples consisted of contiguous bi-decade blocks of tree-rings with a dry weight of around 180g. Each bi-decade sample, after bleaching and charring, was combusted to carbon dioxide. The carbon dioxide was then converted to acetylene which in turn was polymerised to benzene. The typical yield from a 180gm wood sample was 15cc of liquid benzene for scintillation counting.

In order to preserve some semblance of order in working with a long floating chronology which we knew must cover approximately the period 1000 to 4000 BC, and to avoid confusion once the chronology was finally dated, a computer scale was assigned to the long chronology. Again to avoid confusion this scale increased with decreasing age and was chosen so that even if the chronology were completed right to the present the computer scale would not exceed 9999. As it happened the first block of calibration measurements (Pearson et al 1977) was performed on bi-decades centred on 5810 to 4670 (computer scale). We now know, since the completion of the chronology, that the true 'centre' dates of these bi-decades run from 2090 to 3230 BC. As luck would have it the whole BC chronology runs from 5289 to 1 BC while the computer scale ran from 2612 to 7900 i.e. we arbitrarily chose a scale which when dated had 7900 (computer) = 1 BC and of course 7901 = AD 1.

As has been discussed elsewhere (Pilcher, 1980; 1983) the initial interpretation of the 1977 Belfast high-precision results suggested that the calibration was indeed relatively smooth. This was in line with a number of statistical interpretations of Suess's original calibration measurements (see for example Clark, 1975). However three lines of evidence soon caused this smooth interpretation to be called into question. One was the measurement of sections of calibration where 'wiggles' were clearly in evidence (Stuiver, 1978; De Jong et el. 1979; Pearson, 1980). The second was the infilling of the missing samples from the 1977 section of calibration, two of which fell well outside the linear distribution as published. The third was the discovery, deduced from duplicate measurements within the Belfast Laboratory that the precisions quoted on the 1977 results - "1-sigma precision on all dates is $\pm$ 25 yr". (Pearson et al 1977) - were too high! In fact the precision on the original measurements should have been quoted as $\pm$ 20 yr. (Pearson, 1980). The effect of plotting the original results with this reduced precision was to bring back wiggles albeit less dramatic than those proposed by Suess.

As more high precision calibration results became available, it became clear that wiggles in other parts of the calibration were far greater and had much more serious consequences for the interpretation of radiocarbon dates (see Baillie and Pilcher, 1983). The greater part of the calibration for the last 5 millennia was made available at the 1982 Radiocarbon conference in Seattle (Pearson and Baillie, 1983; Pearson et al 1983).

Since the Seattle meeting the Belfast chronology has been completed and all the remaining radiocarbon samples covering the last 7000 years have been measured. These results were presented at the 1985 Radiocarbon Conference in Trondheim and will be published in a special volume of Radiocarbon.

Acknowledgements

The authors wish to thank J. Hillam, E. Halliday, A. Brown, D. Brown and J. Mallory, and M. Munro for assistance with the dendrochronology. The Natural Environment and Science and Engineering Research Councils have generously supported the research.

REFERENCES

Baillie, M.G.L. 1973. A recently developed Irish tree-ring chronology. Tree-Ring Bulletin 33, pp. 15-28.

Baillie, M.G.L. 1974. A tree-ring chronology for the dating of Irish post-medieval timbers. Ulster Folklife 20, pp. 1-23.

Baillie, M.G.L. 1975. A horizontal mill of the eighth century AD at Drumard, Co. Londonderry. Ulster Journal of Archaeology 38, pp. 25-32.

Baillie, M.G.L. 1977a. The Belfast oak chronology to AD 1001. Tree-Ring Bulletin 37, pp. 1-12.

Baillie, M.G.L. 1977b. Dublin medieval dendrochronology. Tree-Ring Bulletin 37, pp. 13-20.

Baillie, M.G.L. 1978. Dendrochronology for the Irish Sea Province in P. Davey (ed.), Man and Environment in the Isle of Man. BAR British Series, 51, pp. 27-37.

Baillie, M.G.L. 1979a. Some observations on gaps in tree-ring chronologies. Proc. Symposium on Archaeological Sciences (Jan 1978), University of Bradford, pp. 19-32.

Baillie, M.G.L. 1979b. An interim statement on dendrochronology at Belfast. Ulster Journal of Archaeology 42, pp. 72-84.

Baillie, M.G.L. 1980. Dendrochronology; the Irish view. Current Archaeology 7(2), pp. 61-3.

Baillie, M.G.L. 1981. Dendrochronology, the prospects for dating throughout Ireland. In: O'Corrian, D. (Ed.) "Irish Antiquity" Festschrift to M.J. O'Kelly Cork, pp. 3-22.

Baillie, M.G.L. 1982. Tree-Ring Dating and Archaeology. Croom-Helm, London.

Baillie, M.G.L. 1983a. Is there a single British Isles oak tree-ring signal? Proc. 22nd Symposium on Archaeometry (Ed.) Aspinall. A. and Warren, S.E. University of Bradford, pp. 73-82.

Baillie, M.G.L. 1983b. Belfast dendrochronology: the current situation Archaeology, Dendrochronology and the Radiocarbon Calibration Curve (Ed.) Ottaway, B.S. University of Edinburgh Occasional paper No. 9, pp. 15-24.

Baillie, M.G.L. and Pilcher, J.R. 1973. A simple cross-dating program for tree-ring research. Tree-Ring Bulletin 33, pp. 7-14.

Baillie, M.G.L. and Pilcher, J.R. 1983. Some observations on the high-precision calibration of routine dates. Archaeology, Dendrochronology and the Radiocarbon Calibration Curve (Ed.) Ottaway, B.S. University of Edingurgh Occasional Paper No. 9, pp. 51-63.

Baillie, M.G.L., Pilcher, J.R. and Pearson, G.W. 1983. Dendrochronology at Belfast as a background to high-precision calibration. Radiocarbon 25, pp. 171-8.

Becker, B. 1981. A 2350 year South German oak tree-ring chronology. Fundberichte aus Baden-Wurttemburg 6, pp. 369-86.

Becker, B. and Schmidt, B. 1982. Verlangerung der mitteleuropaischen Eichern-Jahrringchronologie in das zweite vorchristliche Jahrtausand (bis 1462 v Chr). Archaeologisches Korrespondenzblatt 12, pp. 101-6.

Clark, R.M. 1975. A calibration curve for radiocarbon dates. Antiquity, 49, pp. 251-263.

De Jong, A.J.M., Mook, W.G. and Becker, B. 1979. Confirmation of the Suess wiggles: 3200-3700 BC. Nature 208, pp. 48-49.

Eckstein, D. and Bauch, J. 1969. Beitrag zu rationalisierung eines dendrochronologishen Verfahrens und zu Analyse siener Aussagesicherheit. Forstwis Centralbl., 88, pp. 230-250.

Ferguson, C.W. 1968. Bristlecone pine; science and esthetics. Science, 159, pp. 839-846.

Ferguson, C.W. 1969. A 7104-year annual tree-ring chronology for bristlecone pine, Pinus aristata, from the White Mountains, California. Tree-Ring Bulletin 29, pp. 1-29.

Fletcher, J. 1977. Tree-ring chronologies for the 6th to 16th centuries for oaks of southern and eastern England. Journal of Archaeological Science 4, pp. 335-352.

Hollstein, E. 1973. Jahrringkurven der Hallstattzeit. Trierer Zeitschrift 36, pp. 37-55.

Hollstein, E. 1980. Mitteleuropaische Eichenchronologie Mainz am Rhein.

McKerrell, M. 1975. Correction procedures for C-14 dates. In: T. Watkins (ed.) Radiocarbon: Calibration and Prehistory Edinburgh, pp. 47-100.

Pearson, G.W. 1979. Precise 14-C measurement by liquid scintillation counting. Radiocarbon 21(1), pp. 1-21.

Pearson, G.W. 1980. High precision radiocarbon dating by liquid scintillation counting applied to radiocarbon timescale measurements. Radiocarbon 22(2), pp. 337-47.

Pearson, G.W. and Baillie, M.G.L. 1983. High-precision C14 measurement of Irish oaks to show the natural atmospheric C14 variations of the AD time period. Radiocarbon 25, pp. 187-196.

Pearson, G.W., Pilcher, J.R., Baillie, M.G.L. and Hillam, J. 1977. Absolute radiocarbon dating using a low altitude European tree-ring calibration. Nature 270, pp. 25-8.

Pearson, G.W., Pilcher, J.R., Baillie, M.G.L. 1983. High-precision C14 measurement of Irish oaks to show the natural C14 variation from 200 BC to 4000 BC. Radiocarbon 25, pp. 179-186.

Pilcher, J.R. 1980. Radiocarbon calibration: recent progress. British Museum Occasional Paper 21, pp. 45-51.

Pilcher, J.R. 1983. Radiocarbon calibration and dendrochronology - an introduction. Archaeology, Dendrochronology and the Radiocarbon Calibration Curve (Ed.) Ottaway, University of Edinburgh Occasional Paper No. 9, pp. 5-14.

Pilcher, J.R. and Baillie, M.G.L. 1980a. Six modern oak chronologies from Ireland. Tree-Ring Bulletin 40, pp. 23-34.

Pilcher, J.R. and Baillie, M.G.L. 1980b. Eight modern oak chronologies from England and Scotland. Tree-Ring Bulletin 40, pp. 45-58.

Pilcher, J.R., Hillam, J., Baillie, M.G.L. and Pearson, G.W. 1977. A long sub-fossil tree-ring chronology from the north of Ireland. New Phytologist 79, pp. 713-29.

Pilcher, J.R., Baillie, M.G.L., Schmidt, B. and Becker, B. 1984. A 7272-year European tree-ring chronology. Nature 312, pp. 150-152.

Renfrew, C. and Clark, R.M. 1974. Problems of the radiocarbon calendar and its calibration. Archaeometry 16, pp. 5-18.

Schmidt, B. 1981. Beitrag zum aufbau der holozanen Eichenchronologie in Mitteleuropa. Archaologisches Korrespondenzblatt 11, pp. 361-363.

Schmidt, B. and Schwabedissen, H. 1982. Ausbau des Mitteleuropaischen Eichenjarrhing-kalendars bis in Neolithische Zeit (2061 v. chr.). Archaologisches Korrespondenzblatt 12, pp. 107-108.

Siebenlist-Kerner, V. 1978. The chronology for certain hillside oaks from western England and Wales. BAR International Series, 51, pp. 157-161.

Smith, A.G., Baillie, M.G.L., Hillam, J., Pilcher, J.R. and Pearson, G.R. 1972. Dendrochronological work in progress in Belfast: The prospect for an Irish post-glacial tree-ring sequence. Proc. 8th Int. Conf. on Radiocarbon Dating I, pp. A92-5.

Stuiver, M. 1978. Radiocarbon timescale tested against magnetic and other dating methods. Nature 273, pp. 271-274.

Suess, H.E. 1965. Secular variations of the cosmic-ray-produced C14 in the atmosphere and their interpretations. Journal of Geophysical Research 70, pp. 5937-5950.

Suess, H.E. 1970. Bristlecone pine calibration of the radiocarbon timescale from 5200 BC to the present. In: I.U. Olsson (ed.) Radiocarbon Variation and absolute chronology, Almquist and Wiksell, Stockholm, pp. 303-312.

Dendrochronological Studies of Bog Pine from the
Rannoch Moor Area, Western Scotland

R.G.W.Ward,* B.A. Haggart,[+] and M.C. Bridge[+]

[+]Quaternary Research Unit
Geography Department
City of London Polytechnic
Old Castle Street
London E1 7NT

*Department of Environmental and Geographical Studies
Roehampton Institute
Southlands College
Queensmere Road
London SW 15

ABSTRACT

The paper discusses the potential of sub-fossil Scots pine for chronology building and palaeo-climatic reconstructions in the Rannoch Moor area of Scotland. Although the problems of generally short ring sequences, locally absent rings, greatly compressed growth and lobate growth hinder attempts to build chronologies, sufficient material exists to offer scope for environmental reconstructions. Pollen analyses show that pine were probably present in the Rannoch area continuously at low altitudes from ca 6500 BP to ca 3000 BP, and occurred at higher altitudes in two distinct phases within this period.

INTRODUCTION

The remains of subfossil Pinus sylvestris are often preserved within peat deposits and have been widely reported throughout Scotland including the Outer Hebrides (Wilkins, 1984), the northern Highlands (Lewis, 1905; 1906; Birks, 1975), the Cairngorms (Birks, 1975; Pears, 1975), the western Highlands (Lewis, 1907; Birks, 1975) and the Southern Uplands (Birks, 1975). In a recent study of the distribution of Pinus sylvestris, Bennett (1984) records that the tree reached its maximal geographical extent in Scotland between 7500 and 4000 B.P.

For several years the Quaternary Research Unit at the City of London Polytechnic has been involved in studies of the ecological changes that occurred during the transition between the Late Devensian and Holocene in the Rannoch Moor and surrounding areas (Lowe and Walker, 1981; Walker and Lowe, 1977; 1979; 1980; 1981). At the start of 1981 attention was directed to the subfossil bog pines that are common in the area, in particular to see what scope they offered for chronology building and climatic reconstuction. This paper briefly describes some of the findings from these studies.

The Rannoch Moor Area

Rannoch Moor is a large wilderness covering an area of 500km² at an altitude of between 300-400m OD some 30-40km south-east of Fort William (Figure 1). The surface of the moor is a chaotic landscape of mounds and ridges of glacial debris, separated by irregular depressions containing small lochans or peat bogs. In many places the peat has been gullied by small streams, or exposed in the banks of the small lochans and often many pine stumps can be seen in such locations. These remains are so ubiquitous as to suggest that mature pine forest covered large areas of the Rannoch plateau.

The moor lies close to the boundary of two potential forest zones; predominantly pine forest with some birch and oak towards the north and east and predominantly oak forest with birch to the south and west (McVean and Ratcliffe, 1962). Remains of bog pine in the Rannoch Moor area appear to form a discrete grouping towards the southern limit of the known distribution of subfossil pine in the northern Highlands of Scotland (Birks, 1975; Bennett, 1984). This suggests that during the Holocene pine may have been close to both competitive and autogenic thresholds in the Rannoch Moor area.

Samples of pine stumps have been taken from two locations on the fringes of the moor, selected for their accessibility and the abundance of pine remains. At Clashgour (NN 25 42, altitude ca. 180m OD) a large area of peat, usually between 2 and 3m thick, is exposed on a gently-sloping terrace feature above the modern floodplain of the Allt Linne nam Beathach (Figure 1). Most of the wood occurs as *in situ* stumps, leaving a few centimetres of exposed wood above the highest root. Occasionally large trunks are found which can attain 7m in length and 1.3m in circumference. Many trunks are orientated SW - NE suggesting that they may have been blown down by south-westerly winds. Stumps and trunks lying towards the base of the peat tend to be far better preserved than those higher in the stratigraphy, presumably because permanent waterlogging has restricted the processes of decay. At this site pine does not appear to be confined to distinct stratigraphic horizons.

At Coire Seilich (NN 34 46, altitude ca. 300m OD) there is an area of eroded peat that has formed in the valley of the Allt Coire Seilich, a tributary of the Water of Tulla (Figure 1). Compared to the Clashgour site, the remains of pine are less abundant though better preserved and the peat reaches a depth of ca. 5m.

Scope for Dendrochronology

Approximately 200 samples have been removed from Clashgour and Coire Seilich and ring widths have been measured on one to three radii per sample. Baillie and Pilcher (this volume) underline the necessity for secure crossdating in prehistoric timbers where there is no opportunity for independent verification. This requires sequences of adequate length both to crossmatch with other samples and to extend the chronology. Initial results are not immediately encouraging; only 8% of the samples have more than 200 rings, whilst over 50% have less than 100 rings. Some 90% of the samples may therefore be regarded as of little use in initial chronology building.

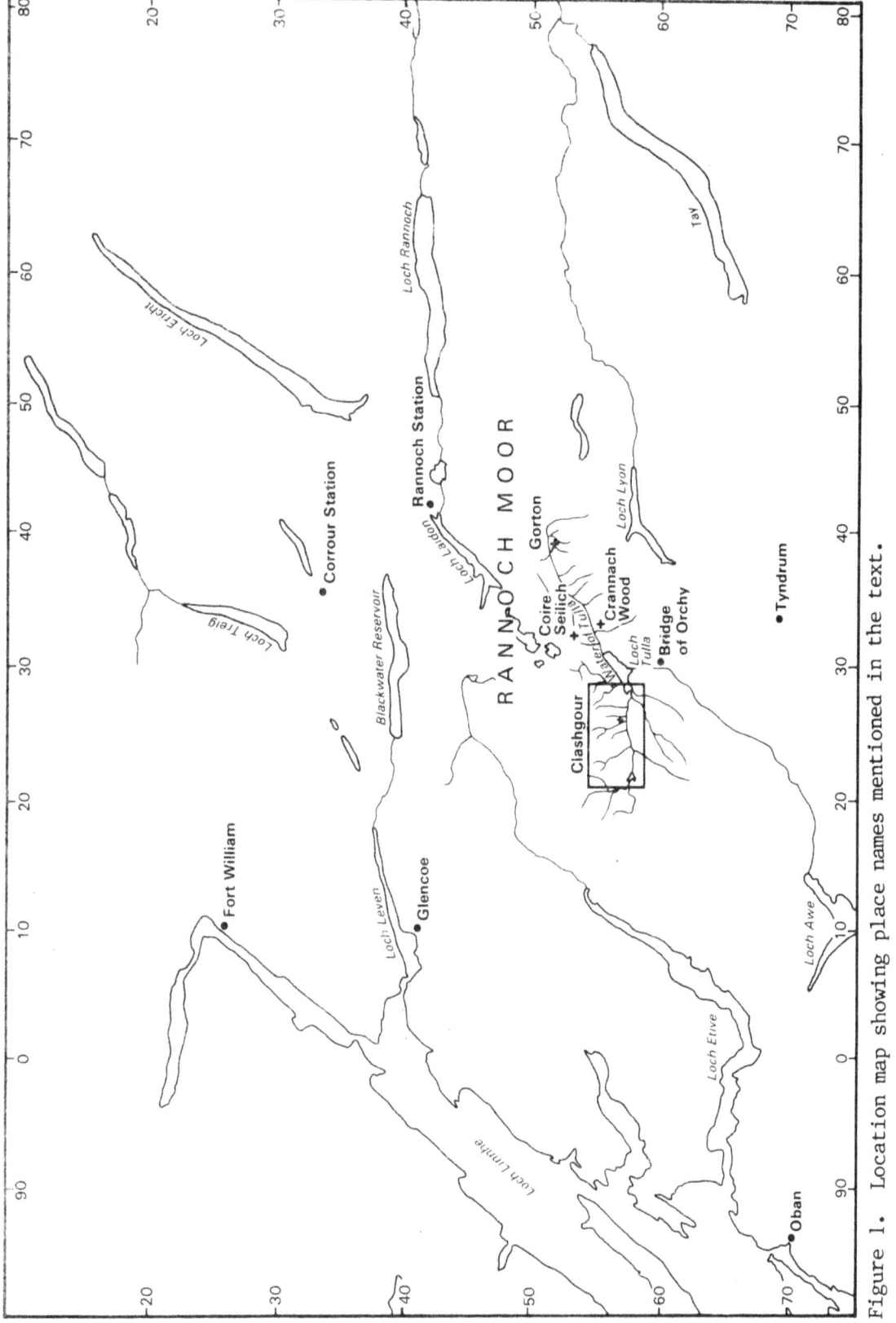

Figure 1. Location map showing place names mentioned in the text.

A second problem with the material collected is the degree to which the cross-sections are affected by both locally missing rings (i.e. those visible around only part of the circumference) and totally missing rings. The former phenomenon may exceptionally affect up to 25% of all rings. Hence comparison of several radii is necessary to obtain an entire ring sequence for each sample and this is often not possible because of damage to the cross-section. It could also be argued that trees with many locally missing rings may be unsuitable for dendrochronological study since the climatic signal may be overidden by other factors affecting growth. Furthermore in view of the high percentage of locally missing rings it is possible that some cross-sections will entirely lack one or more rings, further increasing the difficulty of crossmatching. The longest sequences are associated with slow growing trees and these are more likely to exhibit the problem of locally and totally missing rings.

A third problem is that the ring width can become extremely narrow, involving a thickness of only two or three cells and these are extremely difficult to measure with accuracy. Within such a sequence any year to year variation is compressed so that only a uniform pattern of very narrow rings is left. Again this difficulty is compounded by locally missing rings and is particularly common in the longer ring sequences.

A final problem arises from the variability of a single ring around the circumference of a tree which may be wide in one part and very narrow in another. Sometimes this arises from the proximity of the root zone but it also occurs in sections from the middle of trunks. Values of Student's t for two cross-dated radii from the same tree obtained using the CROS program (Baillie & Pilcher 1973) will frequently fall below 1 as a result of this lobate growth.

Not every sample necessarily suffers from any or all of these difficulties but on average only 2 or 3 samples out of any 100 may be suitable for cross-dating and chronology building. This means that extensive collecting is necessary, involving many sites, to ensure an adequate amount of material is available.

If suitable tree samples can be found, there remains the problem of the 'acceptance level' at which a visual and statistical correlation is sufficiently clear to regard two samples as properly cross-dated. Using modern trees from a variety of locations in northern Scotland, Robertson (1984) found that some good correlations could be obtained, although Student's t rarely reached above 5. In addition, there is a strong temptation to infer missing rings on the basis of visual matches since it is known that locally absent rings are common and because the judicious placement of an extra 'dummy' ring in one sample can make the t statistic jump by 2 or 3! This suggests that a clear rationale must be developed if unambiguous cross-dating is to be achieved.

PROSPECT

One possible technique for improving the usefulness of a larger proportion of samples for chronology building is the measurement of maximum latewood density. Hughes et al. (1984) have shown that far better correlations may be obtained between density measurements than ring widths. However, sub-fossil wood presents the problem of post-mortem changes in the period during which the tree remained buried. Caroline Swain of Liverpool

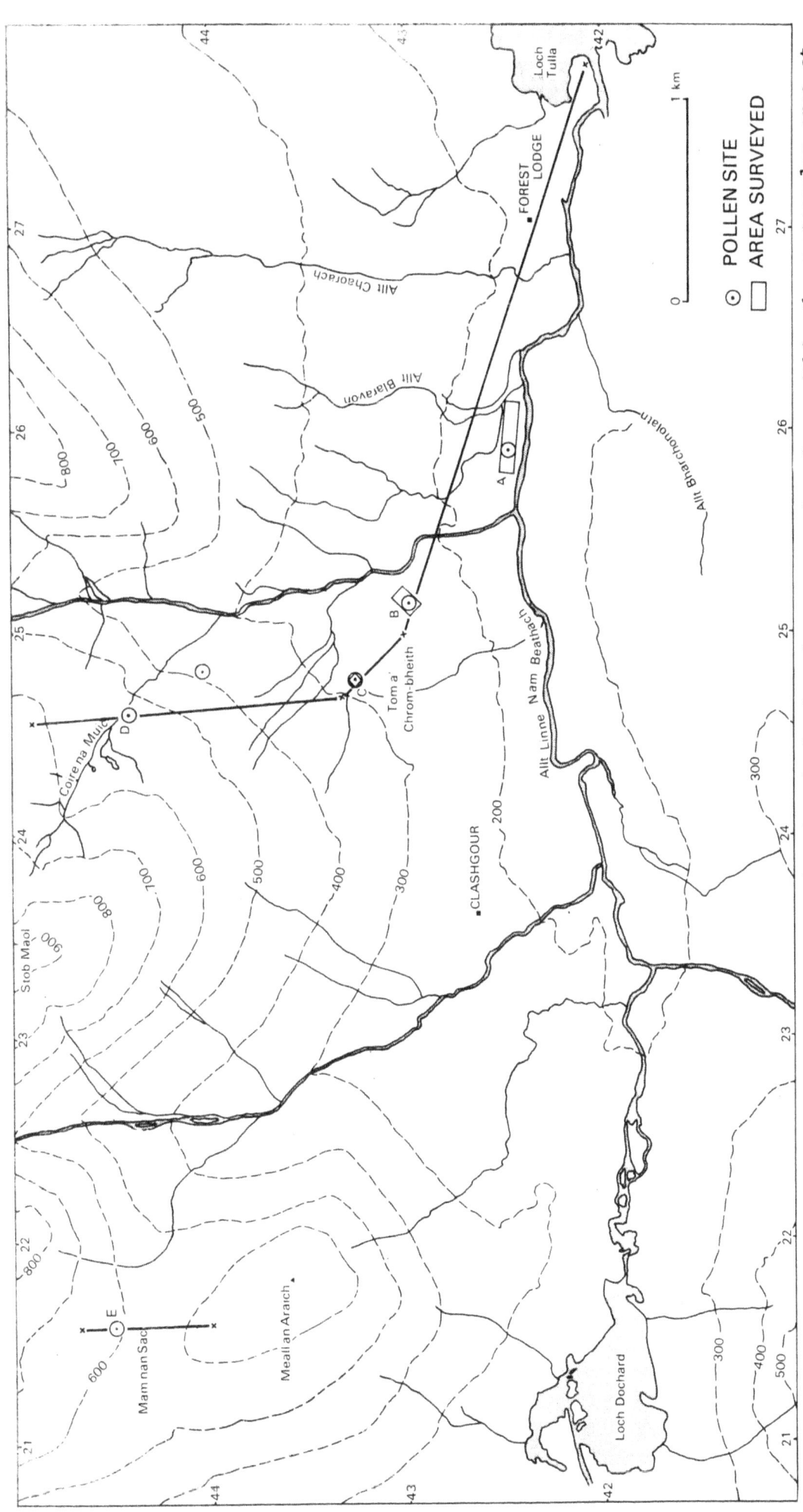

Figure 2. Location map of the Clashgour area. Samples for dendrochronology were taken from within the rectangular area at Clashgour A. The location of pollen sites and the transect used to construct Figure 5 are also shown.

Polytechnic tested whether or not maximum latewood density could be applied to the Rannoch material using some of the best available samples from Coire Seilich and Clashgour. Her pilot study found that good correlations could not generally be obtained even from two radii of the same tree and she concluded that the technique was unlikely to assist in cross-dating separate trees (Swain, pers. comm.). Since density measurements are more difficult to obtain there is likely to be no advantage gained from pursuing this approach.

The potential for obtaining cross dating and a chronology from subfossil Pinus sylvestris probably rests with finding enough material of a broadly similar age. McNally and Doyle (1984) have shown that it is possible to cross-date subfossil pine stumps and the success of European workers with archaeological material shows that pine chronologies can be constructed given the right circumstances. The solution to the problem may rest on good fortune and a suitable approach. The experience of Pilcher and Baillie (this volume) is instructive in this context. Short-lived trees must be ruthlessly avoided and if adequate samples cannot be found at one site, between-site correlation must be assumed. Most importantly, sites must be examined where the tree remains represent a limited period of forest growth. At a site like Clashgour, it is likely that tree growth was continuous for about 2000 - 3000 years. When the majority of trees have less than 100 rings finding overlaps between a series of samples within such a long time period is extremely difficult.

One potential approach lies in the identification of those areas where the tree remains represent a limited total time period of growth. At Clashgour the pollen stratigraphy of a series of sites was investigated along a profile that rises in altitude from ca. 180m (Clashgour A, the site from which material for dendrochronological studies was sampled) to ca. 515m OD (Clashgour D) with an additional site at ca. 603m OD (Clashgour E) some 3km to the west (Figure 2). The pollen curve for Pinus (% Total Land Pollen) together with approximate radiocarbon ages is shown for each of these sites in Figure 3.

Pine pollen is easily transported and therefore high pine pollen frequencies do not necessarily indicate the presence of trees. In order to help interpret the significance of pine pollen frequencies a series of contemporary pollen samples was taken from moss polsters of cushion-forming species in the Crannach Wood area 10km to the east (Figure 1). Crannach Wood is a semi-natural birch-pine woodland fringed at higher altitude by open birch woodland with a ground flora dominated by Molinia, Nardus and sedges. Samples were taken at intervals along a transect from within the birch-pine woodland to the altitudinal limit of the open birch woodland. Within the birch-pine wood pine pollen frequencies dominated and ranged between 76% and 91% TLP (n=4). The average pine frequency for the open birch woodland was 43% (n=9). However, this figure masks large local variations and the range of values was between 7% and 70%.

There are several difficulties in relating these results directly to fossil pine assemblages. First, of prime importance must be the contribution to the pollen rain of vegetation local to the sampling site. In a separate transect to the north-east of the wood where Calluna and Erica rather than grasses and sedges dominate the ground flora, pine frequencies were consistently as low as 1% TLP only 100m from the woodland edge. Therefore fossil pine percentage frequencies must be assessed with respect to the contemporaneous vegetation. Second, the pine pollen within the

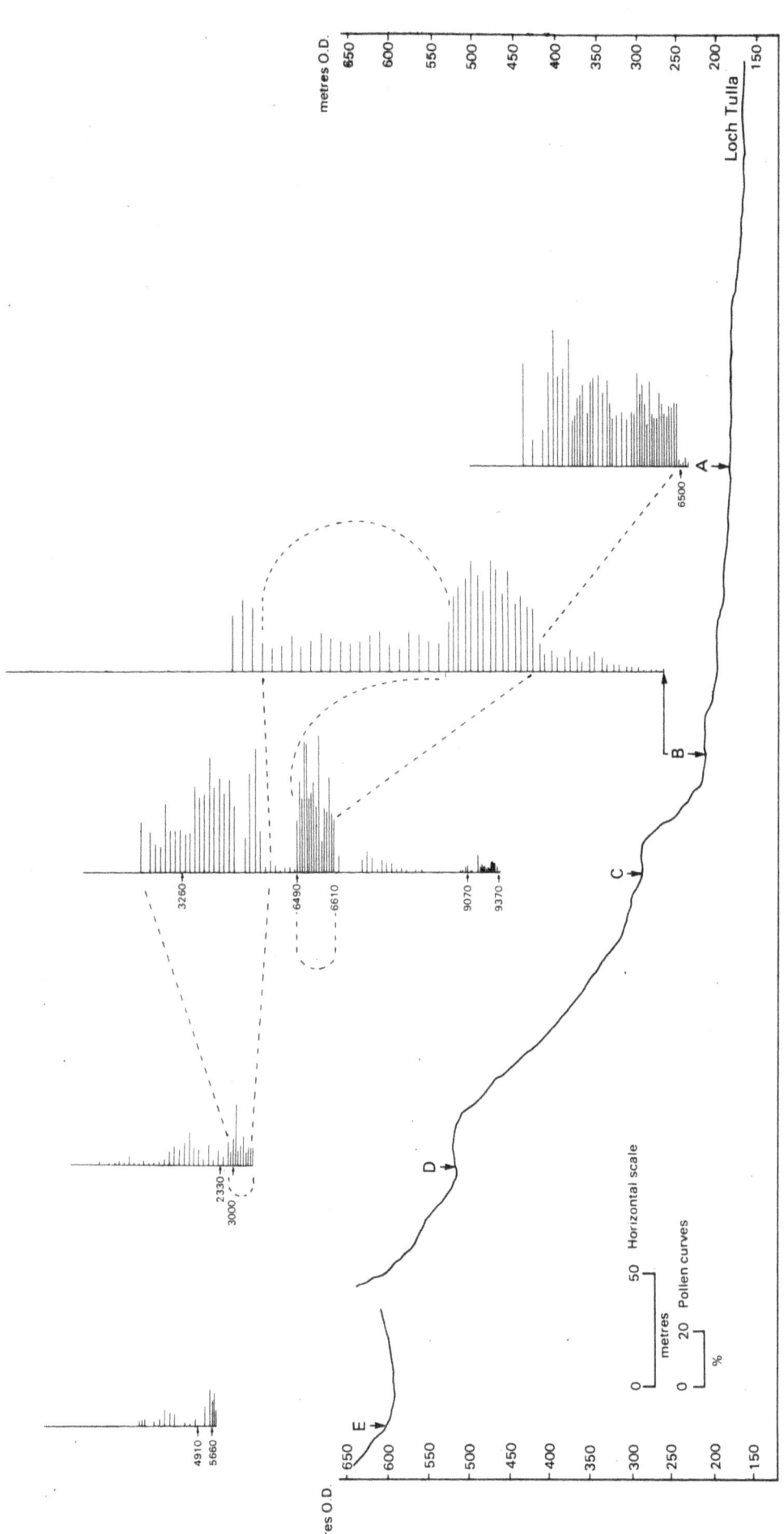

CLASHGOUR

Figure 3. Diagram showing changes in the percentage frequency of Pinus pollen (%Total Land Pollen) with altitude and age in the Clashgour area. All age estimates are based on radiocarbon determinates. The Calshgour B site is euqivalent to Clashgour 2 (Walker and Lowe, 1981) and this pine curve is drawn from their data by kind permission/

present day samples is derived from mature pine trees (there is little evidence of natural regeneration) of up to 250 years age rooted in minerogenic deposits on a well-drained sloping site. In contrast, the trees at Clashgour were all younger samples, the majority rooted directly in peat. Finally the contemporary pollen samples were derived solely from cushion-forming moss species and may represent several years of accumulation. The fossil pollen from Clashgour was derived mainly from herbaceous peat containing a smaller component of mosses which accumulated over an unknown time period. The value of this contemporary pollen study to aid reconstruction of former pine forest limits is therefore greatly restricted by these difficulties.

A different method of reconstructing former pine forest limits involves the direct comparison of fossil pine pollen frequencies with the occurrence of pine macrofossil remains. At Clashgour pine stumps are abundant in the stratigraphic column. Within sites A and C (Figure 2) pine pollen frequencies from herbaceous peat samples next to pine macrofossil remains gave values of over 20% TLP. It seems reasonable to assume, therefore, that figures over 20% TLP derived from herbaceous peat samples can be taken to imply the growth of pine trees adjacent to the sampling site. Indeed, Bennett (1984) in a survey of the Holocene distribution of Pinus sylvestris in Scotland suggests a frequency of 20% Total Pollen may indicate the local presence of the tree.

If this approach is valid, it suggests that pine forest was present at Clashgour A (ca. 180m OD) by 6500 BP. No radiocarbon dates are available for the top of this profile but it is likely that pine remained dominant in the local flora until after the disappearance of elm from the regional vegetation which suggests there may have been continuous presence of pine for between 2000 - 3000 years at this site. At Clashgour B (ca.218m OD) and Clashgour C (ca.293m OD) however there appear to be 2 phases of pine forest growth (Figure 3). Pine began to grow at the latter site from ca.6600 BP and was briefly replaced by birch forest at 6500 BP. This suggests the first phase of pine forest may only have lasted a few hundred years. A second period of pine forest growth began after this date and lasted until about 3000 BP. Pine stump remains extend to ca. 515m OD, the approximate altitude of Clashgour D, but none were found in the broad col of Mam nan Sac adjacent to Clashgour E at ca. 603m OD (Figure 2). It seems likely that the maximum altitudinal extension of pine forest took place during the second phase represented at Clashgour B and C, between ca. 6000 and 3000 BP (Figure 3).

A dendrochronological research programme that is initially aided by pollen stratigraphy and radiocarbon dating appears to offer the best prospects of yielding results with subfossil Pinus sylvestris in western Scotland. At Clashgour attention could have been turned from the lower sites where pine was dominant for between 2000 - 3000 years to those sites at higher altitudes where pine forest growth was at the limit of its range. Pine trees at these sites would have been more susceptible to climatic change and represent a shorter period of growth. However these higher sites are less accessible and contain fewer exposed peat faces.

To gain sufficient samples from the short-lived pine phase between 6600 and 6500 BP at Clashgour B and C would have meant stripping extensive areas of peat to 2 - 3m depth.

Besides dendrochronological information, it is hoped that the bog pines of Scotland may yield further information of ecological significance. If chronology building proves successful, information on the age structure, density and regeneration of these ancient woodlands will be available. Even if chronology building fails, chemical and physical analyses of the wood and their comparison with modern analogues may yield valuable information concerning the environment prevailing during wood formation, and provide evidence which will enable the cause of woodland extinction to be assessed at specific sites. A third site near Gorton in the valley of the Allt Learg Mheuran (NN 39 47, altitude ca. 335m OD, Figure 1) is currently being investigated and information concerning the frequency and basal area of trunks, internode distances and root-mass sizes is being compiled before sampling for dendrochronology.

Acknowledgements

The authors wish to thank Dr. J.J. Lowe for providing useful comments on an earlier draft of this paper, and Miss. C. Swain for help with densitometry. Assistance with fieldwork was given by D. Blake, M. Butler, C. Burgues, Dr. R. Cornish, G. Coston, E.A. Haggart, Dr. P.D. Hulme, Dr. J.J. Lowe, Dr. R.M. Tipping and Dr. D.G. Sutherland. Maps and diagrams were drawn and photographically reduced by L. Judge. Finally we would like to thank Mr. R. Fleming of Black Mount for permitting access to the sites.

REFERENCES

Baillie, M.G.L. and Pilcher, J.R. 1973. A simple cross-dating program for tree-ring research. Tree-ring Bulletin 33, pp. 7-14.

Baillie, M.G.L. and Pilcher, J.R. 1987 (this volume) The Belast Long Chronology Project. pp. 203-214.

Bennett, K.D. 1984. The post-glacial history of Pinus sylvestris in the British Isles. Quaternary Science Reviews 3, pp. 133-155.

Birks, H.H. 1975. Studies in the vegetational history of Scotland. IV. Pine stumps in Scottish blanket peats. Phil. Trans. R. Soc. Lond. B, 270, pp. 181-226.

Hughes, M.K., Schweingruber, F.W., Cartwright, D. and Kelly, P.M. 1984. July-August temperature at Edinburgh between 1721 and 1975 from tree-ring density and width data. Nature 308, pp. 341-344.

Lewis, F.J. 1905. The plant remains in the Scottish peat mosses. Part I. The Scottish Southern Uplands. Trans. R. Soc. Edinb. 46, pp. 33-70.

Lewis, F.J. 1906. The plant remains in the Scottish peat mosses. Part II. The Scottish Highlands. Trans. R. Soc. Edinb. 45, pp. 335-360.

Lewis, F.J. 1907. The plant remains in the Scottish peat mosses. Part III. The Scottish Highlands and the Shetland Islands. Trans. R. Soc. Edinb. 46, pp. 33-70.

Lowe, J.J. and Walker, M.J.C. 1981. The early postglacial environment of Scotland: evidence from a site near Tyndrum, Perthshire. Boreas 10, pp. 291-294.

McNally, A. and Doyle, G.J. 1984. A study of subfossil pine layers in a raised bog complex in the Irish Midlands - 1. Palaeowoodland extent and dynamics. Proc. R. Ir. Acad. 84B, pp. 57-70.

McVean, D.N. and Ratcliffe, D.A. 1962. Plant communities of the Scottish Highlands. HMSO.

Pears, N.V. 1975. Tree stumps in the Scottish hill peats. Scottish Forestry 29, pp. 255-259.

Robertson, J.A. 1984. The suitability of Pinus sylvestris for dendrochronology in western Scotland. Unpubl. MSc. thesis, City of London Polytechnic.

Walker, M.J.C. and Lowe, J.J. 1977. Postglacial environmental history of Rannoch Moor, Scotland, I. Three pollen diagrams from the Kingshouse area. J. Biogeog, 4, pp. 333-351.

Walker, M.J.C. and Lowe, J.J. 1979. Postglacial environmental history of Rannoch Moor, Scotland, II. Pollen diagrams and radiocarbon dates from the Rannoch Station and Corrour areas. J. Biogeog. 6, pp. 349-362.

Walker, M.J.C. and Lowe, J.J. 1980. Pollen analysis, radiocarbon dates and the deglaciation of Rannoch Moor, Scotland, following the Loch Lomond Advance. In Cullingford, R.A., Davidson, D.A. and Lewin, J. (eds.). Timescales in Geomorphology, pp. 247-259, Wiley.

Walker, M.J.C. and Lowe, J.J. 1981. Postglacial environmental history of Rannoch Moor, Scotland, III. Early- and mid-Flandrian pollen stratigraphic data from sites on western Rannoch Moor and near Fort William. J. Biogeog. 8, pp. 475-491.

Wilkins, D.A. 1984. The Flandrian woods of Lewis (Scotland). J. Ecol. 72, pp. 251-258.

The Dendrochronological Study of Sub-Fossil Wood in New Zealand

M.C. Bridge

Quaternary Research Unit
Geography Department
City of London Polytechnic
Old Castle Street
London E1 7NT

ABSTRACT

Recent investigations into sub-fossil remains of Agathis australis for the purpose of chronology construction are briefly outlined. A floating 491 year chronology demonstrates the potential of this material which is shown by radiocarbon analysis to be common throughout the last 40000 years.

INTRODUCTION

Despite New Zealand's small land area, the wide range of habitats and varied flora make it potentially one of the most important countries in the Southern Hemisphere for dendrochronological studies. The development of dendrochronology in New Zealand, from early dating studies up to more recent dendroclimatic investigations, is well charted by Dunwiddie (1979) and Ogden (1982). Although, as they record, many of the early investigations suggested that chronology development would not be possible in many species, Dunwiddie (1979) was able to produce modern chronologies for 7 species out of the 20 he sampled. The extension of these chronologies back in time should be possible and would yield valuable information for a variety of disciplines. Dated wood material from Northern Hemisphere species such as the bristlecone pine (Pinus aristata) in North America and oak (Quercus spp.) in Europe has been used for radiocarbon calibration, studies of the fluctuation in isotopic ratios in wood components and as a basis for palaeoclimatic reconstructions (Fritts 1976; Hughes et al. 1982). Furthermore on a more local scale, sub-fossil wood chronologies have supplied data on the frequency and dating of catastrophic events such as volcanic eruptions and flooding, and longer term geomorphic, hydrologic and vegetational change (Schroder 1980). Dendrochronology could provide a similar database for the Southern Hemisphere, which would be particularly valuable in a country like New Zealand which is geomorphologically active and climatically varied. Already Wilson and Grinsted (1976) have used a single 1000 year old tree in an attempt to derive a palaeotemperature record.

DENDROCHRONOLOGICAL STUDIES OF KAURI (AGATHIS AUSTRALIS)

The allocation of limited resources to the formation of chronologies from sub-fossil wood obviously has to be directed towards studies involving species which are not only generally suitable for dendrochronology, but also for which abundant material from an exensive period is available. Of the species investigated by Dunwiddie (1979) the obvious choice for the formation of long historical chronologies was kauri (Agathis australis). Kauri is a member of the Araucariaceae and is indigenous to New Zealand. Juvenile and adult stages exhibit very different growth forms. The monopodial conical juvenile with a tapering bole gives way to the adult form which has a barely tapering columnar trunk, usually between 12 and 30 m high, supporting a more or less flat-topped spreading crown. It grows naturally on the less fertile soils to a latitude of 38° S although planted examples survive much further south. Individual specimens of this species may live for more than 1000 years (Ecroyd 1982), and a pilot radiocarbon study of bog kauri has shown that material is available throughout the period from the present back to 40000 years BP (Hendy pers. comm.). It was by far the most important tree economically in the early years of European settlement, when its multipurpose timber and gum were exploited on a vast scale.

Studies of modern Kauri

The dendrochronology research group at Auckland University, headed by John Ogden, has concentrated its effort on the investigation of kauri. So far the emphasis has been on gathering information on living trees, with the formation of site chronologies and a statistical study of correlation between appropriate climatic factors and growth, together with more broad-based ecological investigations (Palmer 1982; Palmer and Ogden 1983; Ahmed 1984, and Fowler 1984). The formation of these chronologies has highlighted some of the difficulties of using this species, particularly that it exhibits locally absent rings and pronounced lobate growth. Crossmatching is often very time consuming as only cores are available in most instances.

Most of the modern kauri sites are on drier slopes and ridges, and the chronologies typically have mean ring-widths in the range 1.3 to 1.9 mm, with mean sensitivities in the range 0.20 to 0.35 (Palmer 1982; Ahmed 1984). The response functions (Ahmed 1984; Fowler 1984) show that narrow rings in kauri from these sites are statistically correlated with wet soil conditions.

Studies of Bog Kauri

With an increasing store of knowledge gained from studies of living kauri, the Auckland group widened their investigation to include sub-fossil material.

Sampling was directed towards material at the more recent end of the available time span on the basis that this material was more likely to be matched with living material in the near future. A radiocarbon dating survey of the sub-fossil wood in the Waikato region (Hendy pers. comm.) suggested that much of the material found near Huntly in the Waikato valley north of Hamilton (North Island, Figure 1) was dated to around 3000 BP. Dunwiddie (1979) had previously collected material from this area, but was unsuccessful in crossmatching it. He did however record the observation that many of the sub-fossil trees he encountered ceased growth at the same time of year (early in the growing season), and it was hoped that this may represent a single catastrophe in the area such that many sub-fossil remains would be coeval. This, together with the fact that numerous large remains

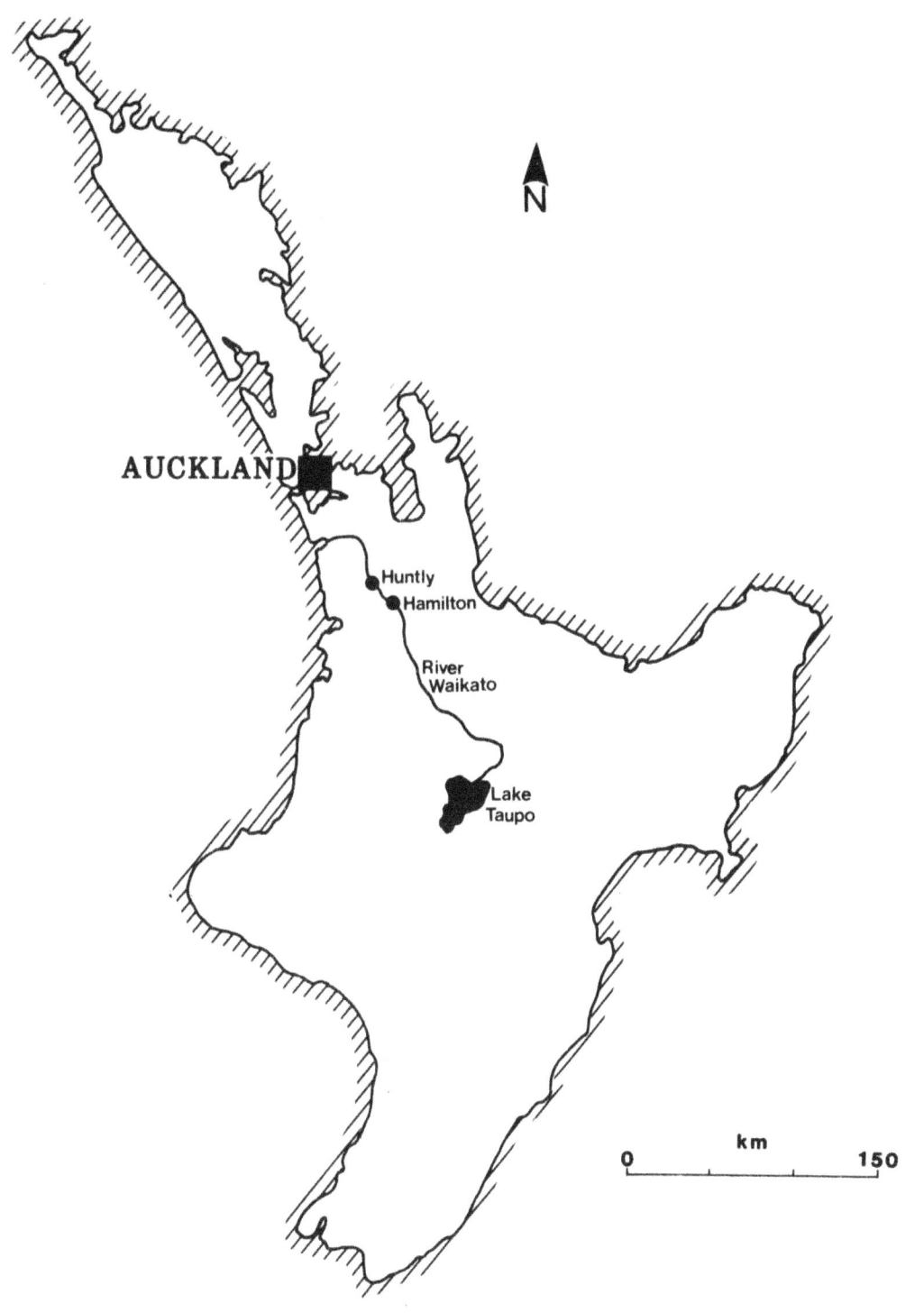

Figure 1. North Island, New Zealand.

up to 1.2m diameter could be found within single "paddocks", and the regions accessibility to Auckland, made the Huntly area the initial focus of investigation (Bridge and Ogden 1986). When sampling was carried out during early 1983, remarkably fresh looking leaves and cones were found buried a few centimetres below the present ground surface. These had been exposed during mechanical ditching work. This reinforced the hypothesis that a single event had led to the demise of at least some of the trees on the site, and that they had been quickly buried.

The samples were transported to Auckland where they were surfaced and examined (Bridge and Ogden 1986). Where possible three radii on each section were measured, but on partial cross-sections only two or one radii were measured. Although it was suspected that the samples may be coeval, it was decided that visual crossmatching of the long plotted ring-width sequences in the first instance may waste a substantial amount of time. As a guide to a subsequent visual crossmatching, program CROS (Baillie and Pilcher 1973) was used in the first instance to identify possible matching positions between trees. This program was devised to give an objective value to the level of agreement between two ring-width sequences known to contain no missing or false rings. It proved invaluable in this work however for highlighting possible crossmatching positions, and areas identified by subsequent visual analysis as possibly containing missing or false rings could be re-examined on the sample. More recently programs produced by Fowler (1984), similar to program ALIGN (Kickert et al. 1983), have superceded the use of program CROS for this crossmatching.

In contrast to the living trees, these sites from which sub-fossil kauri wood is available are mostly situated in flat lowland valleys. One might therefore have expected that the growth of the trees would have been less rapid, and that crossmatching even within the same tree would be less readily achieved. In fact the first sub-fossil tree-ring chronology for the Southern Hemisphere was produced in late 1983 (Bridge and Ogden 1986). In line with the dendroclimatological findings from modern trees this sub-fossil chronology from a presumably damper river valley site has a lower mean ring-width (0.845mm) but a higher mean sensitivity (0.423) than the modern chronologies. Nevertheless the lack of modern analogues for kauri growth on such sites underlines the need for caution in any future palaeoclimatic interpretation derived from this type of material.

The floating 491 year sub-fossil chronology can be approximately dated by radiocarbon analysis to within the period 3500 to 3000 years BP. Whilst this first sub-fossil chronology represents only 5 trees, it shows the potential for future work in the vicinity. Pollen analysis at a bog site within 10km of the sampling site (McGlone et al. 1984) suggest that kauri was present continuously from 7000 BP until cleared in recent centuries, but that it was most abundant in the period 3000 to 2000 years BP. The gathering of material from this district should be a high priority since this valuable source of information is being destroyed rapidly due to large scale drainage and clearance of the land to produce grazing. Material from a second site has already been collected, and preliminary investigations suggest that this too will yield a chronology.

THE FUTURE

Recently (Ogden pers. comm.) the discovery of a 'buried forest' at Pureora, under 2m of peat and pumice from a previous Taupo eruption (central

North Island), has prompted sub-fossil dendrochronological studies of Podocarpus totara and Phyllocladus spp. This work is being carried out by Jonathan Palmer at Auckland University and may provide further scope for palaeoenvironmental reconstructions.

The enormous potential for sub-fossil dendrochronological studies in New Zealand, providing data of local, national and international significance, especially when viewed in the context of the Southern Hemisphere (Ogden 1982), is likely to make this one of the foremost countries for such studies in the near future. The rate of destruction of the source material in areas such as the Waikato valley make the collection and storage of material for subsequent analysis a matter of some urgency.

Acknowledgements

I wish to thank the Research Grants Committee of Auckland University for financial assistance in New Zealand.

REFERENCES

Ahmed, M. 1984. Ecological and dendrochronological studies on Agathis australis Salisb. (Kauri). PhD thesis, Auckland University.

Baillie, M.G.L. and Pilcher, J.R. 1973. A simple cross-dating program for tree-ring research. Tree-Ring Bulletin 33 pp. 7-14.

Bridge, M.C. and Ogden, J. 1986. A sub-fossil kauri (Agathis australis) tree-ring chronology. Journal of the Royal Society of New Zealand 16 pp. 17-23.

Dunwiddie, P.W. 1979. Dendrochronological studies of indigenous New Zealand trees. New Zealand Journal of Botany 17 pp. 251-266.

Ecroyd, C.E. 1982. Biological Flora of New Zealand: 8 Agathis australis (D. Don) Lindl. (Araucariaceae) Kauri. New Zealand Journal of Botany 20 17-36.

Fowler, A. 1984. A dendrochronological study of kauri (Agathis australis). MSc Thesis. Auckland University.

Fritts, H.C. 1976. Tree Rings and Climate. Academic Press, London.

Hughes, M.K., Kelly, P.M., Pilcher, J.R. and LaMarche, V.C. (eds.) 1982. Climate from Tree Rings. Cambridge University Press.

Kickert, R.N., Herren-Gemmil, B., Arkley, R. and Thompson, R.E. 1983. Curves and Align. Tree-Ring Bulletin 43 pp. 79-88.

McGlone, M.S., Nelson, C.S. and Todd, A. J. 1984. Age, vegetational history and environmental significance of pre-peat and surficial peat deposits at Ohinewai, Lower Waikato Lowland, New Zealand. Journal of the Royal Society of New Zealand 14, pp. 233-244.

Ogden, J. 1982. Australasia: In Hughes, M.K., Kelly, P.M., Pilcher, J.R. and LaMarche, V.C. (eds.) Climate from Tree Rings. pp. 90-103.

Palmer, J. 1982. A dendrochronological study of Kauri. MSc thesis, Auckland University.

Palmer, J. and Ogden, J. 1983. A dendrometer band study of the seasonal pattern of radial increment in Kauri (Agathis australis). New Zealand Journal of Botany 21, pp. 121-126.

Schroder, J.F. Jr. 1980. Dendrogeomorphology: review and new techniques of tree-ring dating. Prog. in Phys. Geog. Vol. 4, No. 2, pp. 161-188.

Wilson, A.T. and Grinsted, M.J. 1976. The possibilities of deriving past climate information from stable isotope studies on tree-rings. In: Proceedings International Conference on Stable Isotopes, pp. 1-12. Institute of Nuclear Science, Lower Hutt, New Zealand.

www.ingramcontent.com/pod-product-compliance
Ingram Content Group UK Ltd.
Pitfield, Milton Keynes, MK11 3LW, UK
UKHW060200240426
12048UKWH00029B/1672